SPECIAL PAPERS IN PALAEONTOLOGY NO. 83

———

SILURIAN CONODONTS FROM THE YANGTZE PLATFORM,
SOUTH CHINA

BY

WANG CHENG-YUAN
and RICHARD J. ALDRIDGE

with 13 text-figures, 30 plates and 9 tables

THE PALAEONTOLOGICAL ASSOCIATION
LONDON

June 2010

CONTENTS

[Special Papers in Palaeontology, 83, 2010, pp. 5–136]

Abstract: Silurian conodonts from several sections in the area of the Yangtze Platform, south China, are described and their taxonomy revised. Two new families, Pseudooneotodidae and Gamachignathidae, are erected, one new genus, *Chenodontos*, and ten new species and subspecies: *Apsidognathus ruginosus scutatus*, *Chenodontos makros*, *Distomodus cathayensis*, *Oulodus tripus*, *Ozarkodina wangzhunia*, *Panderodus amplicostatus*, *Pterospathodus sinensis*, *Wurmiella amplidentata*, *Wurmiella curta* and *Wurmiella recava*. Some additional new taxa are introduced in open nomenclature. The status of Silurian conodont biozonation in China is reviewed, and the following successive appearance biozones provisionally recognised to span the Llandovery succession in ascending order: *Ozarkodina* aff. *hassi* Biozone; *Ozarkodina obesa* Biozone; *Ozarkodina parahassi* Biozone; *Ozarkodina guizhouensis* Biozone; *Pterospathodus eopennatus* Biozone; *Pterospathodus celloni* Biozone; *Pterospathodus amorphognathoides* Biozone. The re-assessment of conodont data indicates that unequivocal Wenlock taxa have not been recorded on the Yangtze Platform and that Wenlock marine deposits, if present, are much less extensive than previously suggested. Current conodont evidence also indicates that red beds are probably developed at three levels in the Silurian of the region: upper Aeronian to lower Telychian; upper Telychian, perhaps extending into the Wenlock; Ludlow, pre-*O. crispa* Biozone.

Key words: conodonts, Silurian, China, systematics, biostratigraphy.

SILURIAN conodonts are abundant and well preserved in many sections in south China. A number of papers have been published recording these faunas and applying them in local biostratigraphy, but the majority are in Chinese and most do not employ multielement concepts in their taxonomy. For these reasons, there has been little integration of the Chinese data with those from other parts of the world. In this article, we describe collections from a number of sections in south China, re-evaluate existing publications and refine the taxonomy of several Silurian species. We also use the conodont information to suggest correlations of Chinese Silurian sections with successions elsewhere and to contribute to the understanding of the palaeoecological, palaeobiogeographical and evolutionary patterns displayed by Silurian conodont faunas.

PREVIOUS RESEARCH ON CHINESE SILURIAN CONODONTS

The study of Silurian conodonts in China is still in its infancy. The first descriptions were published as recently as 1980 by Wang Cheng-yuan, who reported Upper Silurian conodonts, including *Ozarkodina crispa*, from eastern Yunnan. This species was subsequently subdivided into four different morphotypes on the basis of Chinese material (Walliser and Wang 1989; see also Wang C. Y. 2001).

Llandovery conodonts were first described by Zhou *et al.* (1981), who proposed a zonal scheme based on the Leijiatun and Tudiao sections in Guizhou Province. These authors identified 93 form species, referred to 42 form genera; 35 of the form species and seven of the form genera were new. Although this work is in need of taxonomic revision, it established the framework of the Llandovery conodont sequence in south China, with five zones and two subzones defined: '*Spathognathodus obesus*' Assemblage Zone; Interval-Zone 1; '*Spathognathodus parahassi* – *S. guizhouensis*' Assemblage Zone (with lower, '*S. parahassi*', and upper, '*S. guizhouensis*', assemblage subzones); '*Spathognathodus celloni*' Assemblage Zone; Interval-Zone II. Additional documentation of the conodonts from these sections was provided by Zhou and Zhai (1983), who erected a further three form genera and eight form species. Subsequently, Zhou (1986) provided a discussion of environmental controls on Early Silurian conodonts; he also assessed the thermal maturation of Silurian strata using the colours of conodont elements (Zhou 1980, 1983) and summarized the Silurian conodont biostratigraphy of China (Zhou and Yu 1984; Zhou *et al.* 1985).

Llandovery conodonts were also reported by Ni S. Z. (1987), who identified and described 23 form species (four new) assigned to 14 form genera from the Yangtze Gorge Area. He also established three unnamed Assemblage Zones, although he did not define the boundaries.

WANG CHENG-YUAN

Nanjing Institute of Geology and Palaeontology, Chinese Academy of Sciences, Nanjing 210008, China; e-mail: cywang@nigpas.ac.cn

RICHARD J. ALDRIDGE

Department of Geology, University of Leicester, Leicester LE1 7RH, UK; e-mail: ra12@le.ac.uk

The recognition of Silurian conodont biozones in China was extended by the reports of *Pterospathodus celloni*, *P. amorphognathoides*, 'Spathognathodus sagitta bohemica', *Ancoradella ploeckensis*, *Polygnathoides siluricus* and *Ozarkodina remscheidensis eosteinhornensis* biozones in the Himalaya district of Tibet by Lin (1983) and Lin and Qiu (1983), although these authors did not illustrate the faunas. More firmly supported documentation of the *P. amorphognathoides* Biozone from western Yunnan and Tibet was provided by Wang C. Y. and Wang Z. H. (1981) and by Wang C. Y. and Ziegler (1983). Yu (1985) described and illustrated six form species assigned to six form genera and recognized four conodont biozones (those of *P. celloni*, *P. amorphognathoides*, 'Spathognathodus' sagitta and *Polygnathoides siluricus*) in the Xainza area of north Tibet, and Qiu H. R. (1985) illustrated specimens from six conodont biozones (*P. celloni*, *P. amorphognathoides*, 'S. sagitta bohemicus', *A. ploeckensis*, *P. siluricus* and *O. eosteinhornensis*) and two faunas (the *K. variabilis* and *O. excavata excavata* faunas) from the Himalaya district and the Gandise-Nyainqentanglha district of north Tibet. Following additional fieldwork, she further identified the same conodont sequence in Tibet (Qiu H. R. 1988).

Additional records of Llandovery conodonts were provided by Ding and Li (1985), who made the first collections from the important succession in the Ningqiang area, Shaanxi Province. They identified 70 form species assigned to 30 form genera, including 13 new form species and two new form genera. Again, multielement revision of their taxonomy has been necessary, and the re-evaluation undertaken herein means that few of their new names survive. They also named two assemblage subzones within the 'Spathognathodus celloni Assemblage Zone': the 'Aulacognathus bashanensis Sub-assemblage Zone' below and the 'Ozarkodina adiutricis Sub-assemblage Zone' above. However, neither of these names is sustainable; *Aulacognathus bashanensis* is a subjective junior synonym of *Aulacognathus bullatus*, and *Ozarkodina adiutricis* is the name originally given to the Pb element of the zonal species, *Pterospathodus celloni*, of which it is a junior synonym. Therefore, these subdivisions cannot be applied. Telychian conodonts were also reported from north-western Hunan by Zuo (1987), who listed 100 form species and subspecies placed in 30 form genera; these records are unsupported by any description or illustration, and the 20 new species/subspecies names and the one new genus must be disregarded. This list does, however, indicate the presence of the *P. celloni* Biozone in the Wujiayuan Formation of the Dayong-Sangzhi area. Additional collections from western Hunan were documented by Wang G. X. *et al.* (1988).

An (1987) produced a monograph on the Lower Palaeozoic conodonts of south China, in which he described and illustrated 21 Silurian form species assigned to 12 form genera. From these faunas, he identified the occurrence of

four Silurian conodont biozones in south China: *Pterospathodus celloni*, *P. amorphognathoides*, *Ozarkodina sagitta bohemica* and 'Spathognathodus crispus'. Important biostratigraphical data were also published by An and Zheng (1990), who illustrated 12 species including *Pterospathodus posteritenuis* Uyeno and Barnes (= *Pranognathus posteritenuis*) from the upper part of the Zhaohuajing Formation in Tongxin County, Ningxia Autonomous Region. The age of this formation had been disputed for a long time, but on the basis of the conodonts they were able to conclude that it could not be younger than the *Monograptus sedgwickii* chron, assigning it to the late Idwian to early Fronian (middle Aeronian Stage of current terminology).

A Silurian conodont sequence was documented from the Erlangshan area, Sichuan Province, by Yu (*in* Jin *et al.* 1989), whose scheme included three 'assemblage zones', three 'interval zones' and one 'range zone'. In ascending order, these are as follows:

Hamarodus europaeus – *Hadrognathus staurognathoides Interval Zone*

This 'interval zone' is equivalent to the lower part of the Machangpo Formation, in which only a few conodonts were found. Yu (*in* Jin *et al.* 1989) used two species that occurred below and above this interval, respectively, to constrain a zone that lacked them, but this is of little practical value.

Pterospathodus celloni – *Hadrognathus staurognathoides Assemblage Zone*

This zone, covering the upper part of the Machangpo Formation and the lower part of the Luoquanwan Formation, is very broadly conceived and contains representatives of at least two previously established and widely recognized biozones.

Hindeodella equidentata – *Pterospathodus celloni Interval Zone*

This 'interval zone' represents about 260 m of the upper Luoquanwan Formation in which virtually no conodonts were found.

Hindeodella equidentata – *Plectospathodus extensus Range Zone*

This 'range zone' is represented in a succession referred by Yu to the ' Middle Silurian' (= Wenlock), including the Londanyan, Changyanzi and Baohuoyan formations. Among the conodonts identified in these formations are *Aulacognathus bullatus*, *Ozarkodina guizhouensis*, *O. hassi* and *Distomodus staurognathoides*, all known only from the Llandovery. The zonal concept disregards these taxa and is based on elements that belong to the long-ranging multielement species *Wurmiella excavata*.

Spathognathodus crispus – Hindeodella equidentata *Interval Zone*

This 'interval zone' represents the Yanziping Formation, between the Baohuoyan and the Sashuiyan formations, where 279 m of strata have yielded no conodonts at all.

Spathognathodus crispus *Assemblage Zone*

This assemblage zone is represented in the upper part of the Sashuiyan Formation and equates with the standard *Ozarkodina crispa* Biozone (see Aldridge and Schönlaub 1989) of late Ludlow age.

Lonchodina walliseri – Trichonodella inconstans – Ligonodina silurica *Assemblage*

This 'assemblage' represents the conodonts from the Maliuqiao Formation and was assigned by Yu to the Přídolí. The occurrence of *Ozarkodina crispa* in the lower Maliuqiao Formation, however, indicates that at least this part of the formation is probably upper Ludlow; the upper part might be Přídolí. The assemblage name used is based on elements of a species of *Oulodus*, with little demonstrated biostratigraphical value.

Additional information on Llandovery conodonts was provided by Liu *et al.* (1993), who reported faunas from Longmen Mountain in Sichuan Province. These authors followed the taxonomy of Zhou *et al.* (1981) and Ding and Li (1985) and identified a conodont sequence closely similar to that established by the former authors in the Leijiatun section. However, they made a few modifications to the previous zonations, of which the most important were the following: (1) a subdivision of the *Spathognathodus obesus* Zone into a lower Subzone A and an upper Subzone B; (2) the establishment of an 'Unnamed Zone I' between the '*Spathognathodus guizhouensis* – *S. parahassi* Assemblage Zone' of Zhou *et al.* (1981) and the '*Ozarkodina adiutricis* – *Aulacognathus bashanensis* Assemblage Zone' of Ding and Li (1985); (3) the subdivision of this unnamed zone into a lower Subzone C and an upper Subzone D. Unfortunately, no biostratigraphical data were provided to enable recognition

of the unnamed zone or any of the new subzones; in the absence of adequate definitions, these divisions cannot be utilized or correlated with other schemes.

There have also been a number of other reports of Llandovery conodonts in China. A small fauna from the Chenxiacun Formation of Hangshan County in Anhui was reported by Wang C. Y. (1993), and the age of the Yimugantawu Formation in the Tarim basin (previously regarded as Late, Middle or Early Devonian) was shown on the basis of conodonts to be Aeronian to Telychian in age (Zhang and Wang 1995). The lower part of the *P. amorphognathoides* Biozone has been reported using reliable data from Hanjiga Mountain in northern Xinjiang by Xia (1993). The record of the lower part of the *P. amorphognathoides* Biozone in the Guangyuan-Ninqiang area by Qian (*in* Jin *et al.* 1992) is less well established, as the zonal taxon was not recognized; however, one of the specimens she illustrated as '*Spathognathodus pennatus procerus*' (pl. 3, fig. 12a–b) has well-developed platform ledges and might well indicate this zone. The other specimen she assigned to this taxon (pl. 3, fig. 6) is comparable to a plexus of pennate morphotypes that are known to occur near the top of the *P. celloni* Biozone (Männik and Aldridge 1989; Männik 1998) and is here referred to *P. amorphognathoides* aff. *lennarti*. More recently, Li and Qian (2001) reported the *P. celloni* and *P. amorphognathoides* biozones from the Ninglang-Yanbian region on the western margin of the Yangtze Platform, and Jin *et al.* (2005) documented these two zones succeeded by a range of zones through to the Lower Devonian in Yanbian, Sichuan Province. In the latter paper, the presence of Ludlow strata is well attested by the identifications of *Ancoradella ploeckensis* Walliser, *Kockelella variabilis ichnusae* Serpagli and Corradini, *Polygnathoides siluricus* Branson and Mehl, and other taxa. Some of the supposed Wenlock taxa are less secure; the specimen illustrated as *Kockelella patula* Walliser (Jin *et al.* 2005, pl. 1, figs 15–16) looks like a species of the Llandovery genus *Aulacognathus*, and the specimens illustrated as *Ozarkodina sagitta rhenana* (Walliser) and *O. sagitta bohemica* (Walliser) (Jin *et al.* 2005, pl. 1, figs 6–7, 13–14) lack the characteristic flared cavities and lateral profiles of these taxa.

Other papers documenting Upper Silurian conodonts include records of a few taxa from north-west Qiangtang and the Karakorum region (Wang C. Y. 1998). A Ludlow fauna, including *A. ploeckensis* and a new species, *Ozarkodina uncrispa*, was reported by Wang P. (2004, 2005) from Inner Mongolia. A wider variety of Silurian species, including *Icriodella inconstans*, '*Ozarkodina inclinata*', *Ozarkodina crispa*, *Pterospathodus pennatus pennatus*, *P. p. procerus* and *Pseodooneotodus bicornis*, were documented by Wang C. Y. *et al.* (2004) from Xianza (Shenzha) County, north Tibet.

Most recently, a typical *P. eopennatus* Zone Llandovery fauna has been recognized in the Baizitian section, Yanbian County, Sichuan (Wang C. Y. *et al.* 2009), and a similar fauna has been found in the uppermost Shamao Formation at the Yanglin section in Zigui County, Hubei (Wang C. Y. *et al.* 2010).

In 1988, a joint programme was initiated between the Chinese Academy of Sciences and the Royal Society, entitled 'Transhemisphere Telychian: a biostratigraphical experiment'. The purpose of this project was to use a variety of fossils, including conodonts, to correlate Telychian sections in China as precisely as possible with coeval British sequences. The present authors were involved in this exercise and undertook joint fieldwork in China and joint laboratory work in England. Monographs have been published in Chinese and English documenting the results of the Transhemisphere Telychian (TT) project (Chen and Rong 1996; Holland and Bassett 2002), including conodont biostratigraphical data. However, it became clear during the course of the programme that, although the Chinese conodont faunas were very interesting and evolutionarily important, the taxonomy of many of the species was in need of multielement revision and that a number of new taxa required description. The same applies to taxa from the pre-Telychian part of the Llandovery at the important Leijiatun section, which was also sampled during the TT project. This systematic revision is the main objective of this paper.

SECTIONS STUDIED

The conodont collections made specifically for the Transhemisphere Telychian project derive mainly from three sections: the Leijiatun section, Shiqian County, Guizhou Province; the Yushitan section, Ningqiang, Shaanxi Province; and the Xuanhe section, Guangyuan City, Sichuan Province. In addition, conodont samples have been collected from two other sections: the Huanggexi section, Daguan County, Yunnan Province; the Erlangshan section, Tianquan county, Sichuan Province. All the conodonts described in this article are from these five sections (Text-fig. 1), plus one isolated locality at Zheng'an, north Guizhou Province. The Leijiatun section comprises a series of natural exposures and road cuts and displays an almost complete sequence from the uppermost Ordovician to the upper Llandovery Series, Lower Silurian. It is here that Zhou *et al.* (1981) primarily established their local conodont zonation for the Llandovery Series. Conodonts are abundant in many of the formations in the Leijiatun section, and we have adopted it as our reference section for Lower Silurian conodont biostratigraphy in south China. The Huanggexi section at Daguan has been claimed to be a more complete Silurian section, representing the interval from the latest Ordovician to the Ludlow (Lan 1979), but this is not corroborated by the conodont faunas. An extensive section through the Silurian has also been reported from Erlangshan in Sichuan

TEXT-FIG. 1. Outline map of south China and adjacent regions, showing tectonic units and the locations of sampled sections (modified after Chen *et al.* 2002).

Province. Jin *et al.* (1989) published a monograph on the Silurian stratigraphy and palaeontology of the Erlangshan district, in which they concluded that the Erlangshan section displays a continuous sequence of marine deposits, representing all the Silurian series. Of particular importance is their identification of Wenlock marine deposits and fossils, including conodonts. If correct, this would be a unique record of marine Wenlock deposits in south China, but their conodont records seem to indicate a Llandovery age for the relevant strata (see above). To resolve this, a re-investigation of the conodonts from the Erlangshan section has been undertaken.

The five sections are outlined below. More detailed stratigraphical and lithological descriptions of the first three sections were given by Chen *et al.* (2002). For the Huanggexi section, the papers by Lan (1979) and Ye *et al.* (1983) are recommended. The monograph by Jin *et al.* (1989) gives most detail for the Erlangshan section.

The Leijiatun section

Leijiatun is a small village, 8 km north of Shiqian county town, north-eastern Guizhou Province. A nearly continuous sequence from upper Ashgill to upper Llandovery is well exposed along the road from Leijiatun to Dashitou (Text-fig. 2). This section is the type section for the Xiangshuyuan, Leijiatun and Majiaochong formations and is a major reference section for the Llandovery Series in south China (see Chen *et al.* 2002, pp. 30–31); it was

originally measured by Ge *et al.* (1979), and a summary section is shown in Text-figure 3. Range charts of conodonts are given in Text-figures 4 and 5, and details of elements recovered in Tables 1–5 (see Appendix).

The base of the collected section is taken at the Kuanyinchiao Bed, a 1.1-m-thick massive grey limestone, lying disconformably on the shales and nodular limestones of the Linshiang Formation. Graptolites in the beds above suggest that this is of latest Ordovician (Hirnantian) age, but the presence of *Ozarkodina* aff. *hassi* indicates a Silurian aspect to the conodont fauna.

The succeeding Lungmachi Formation comprises 3.3 m of dark grey graptolitic shales. Here, the graptolite fauna is dominated by species of *Climacograptus*, but at the Tongzi section in northern Guizhou, where the formation is 181 m thick, Chen and Lin (1978) identified graptolites of the *persculptus* Biozone in the lowest beds. Impersistent calcareous bands occur at Leijiatun, and two from the base of the formation were sampled for conodonts; these are argillaceous ostracod packstones, with a high phosphate content. These samples yielded no conodont elements, but were extremely rich in ostracods, including some palaeocopid forms.

The Xiangshuyuan Formation comprises 76 m of marls, nodular limestones and beds of bioclastic limestone. The whole formation is rich in brachiopods and corals, with occasional nautiloids. The brachiopod dominated community low in the formation (TT 819) suggests a shallow subtidal environment. Samples throughout the formation

TEXT-FIG. 2. Locality map for the Leijiatun section, Shiqian County, Guizhou Province.

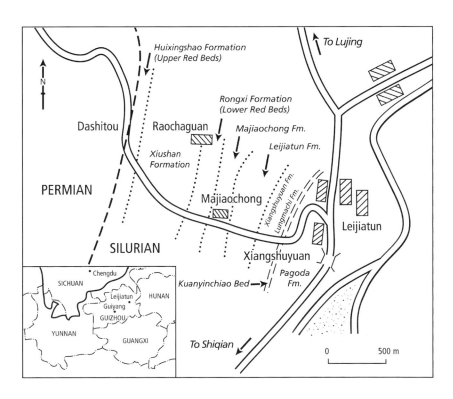

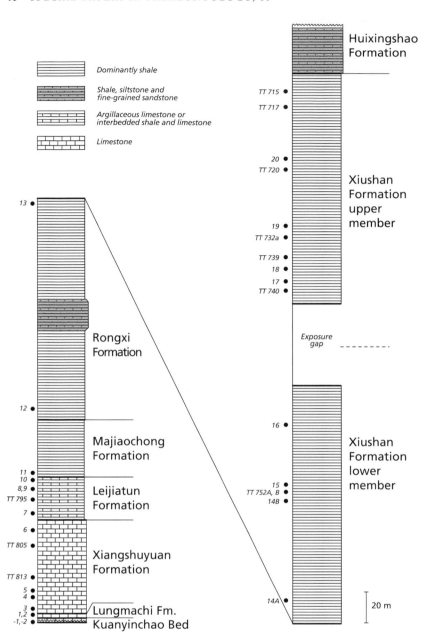

Dominantly shale

Shale, siltstone and fine-grained sandstone

Argillaceous limestone or interbedded shale and limestone

Limestone

Huixingshao Formation

TT 715

TT 717

20
TT 720

Xiushan Formation upper member

19
TT 732a

TT 739
18
17
TT 740

Exposure gap

13

Rongxi Formation

12

Majiaochong Formation

11
10
8,9
TT 795
7

Leijiatun Formation

6

TT 805

Xiangshuyuan Formation

TT 813

5
4

3
1,2
-1,-2

Lungmachi Fm.
Kuanyinchao Bed

16

Xiushan Formation lower member

15
TT 752A, B
14B

14A

20 m

TEXT-FIG. 3. Summary measured section for the Leijiatun section, showing the sample horizons of samples Shiqian-2 to Shiqian 20 (numbers only shown) and of key TT samples.

have yielded conodonts, with the richest fauna coming from an intrasparite in the lower part of the formation (Shiqian 5, = TT 815).

The base of the Leijiatun Formation is taken for this study at the top of the last persistent limestone of the Xiangshuyuan Formation. The Leijiatun Formation is 32 m thick; it consists dominantly of greenish marls and shales, with bioclastic limestones increasing in frequency in the higher part. The bioclastic limestones probably represent storm deposits, often with scoured bases; samples from these throughout the formation have yielded conodonts, although a sample of argillaceous fine-grained limestone (Shiqian 10) near the top of the formation, immediately below a bed with promi-

nent corals and crinoids, produced only a few specimens.

The conformably succeeding Majiaochong Formation comprises 48 m of green-yellow shales without limestone beds or lenses. A low diversity brachiopod fauna has been reported (Chen *et al.* 2002), and the basal part of the formation contains the chitinozoans *Conochitina iklaensis* Nestor and *Ancyrochitina shiqianensis* Geng, indicative of the *sedgwickii* Graptolite Biozone. Samples from this formation have not produced conodont elements.

The Rongxi Formation corresponds to the 'lower marine red beds' of earlier literature. It is 179 m thick and consists of purple-red, grey-green and blue-grey shales

TEXT-FIG. 4. Range chart of biostratigraphically important conodonts from the Kuanyinchiao Bed (K) through to the Leijiatun Formation, Leijiatun section.

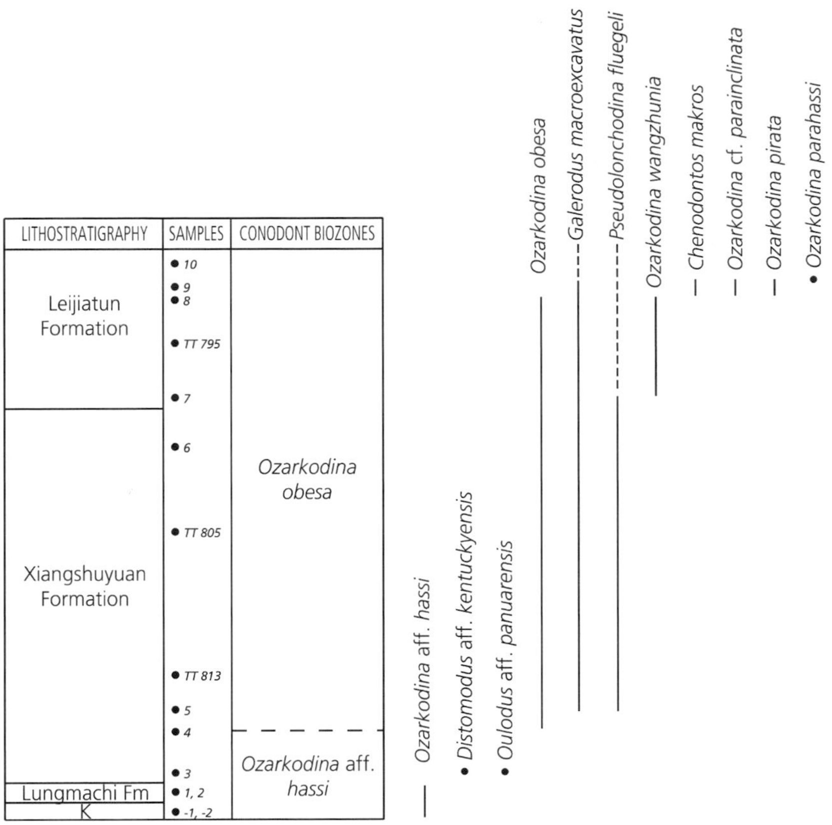

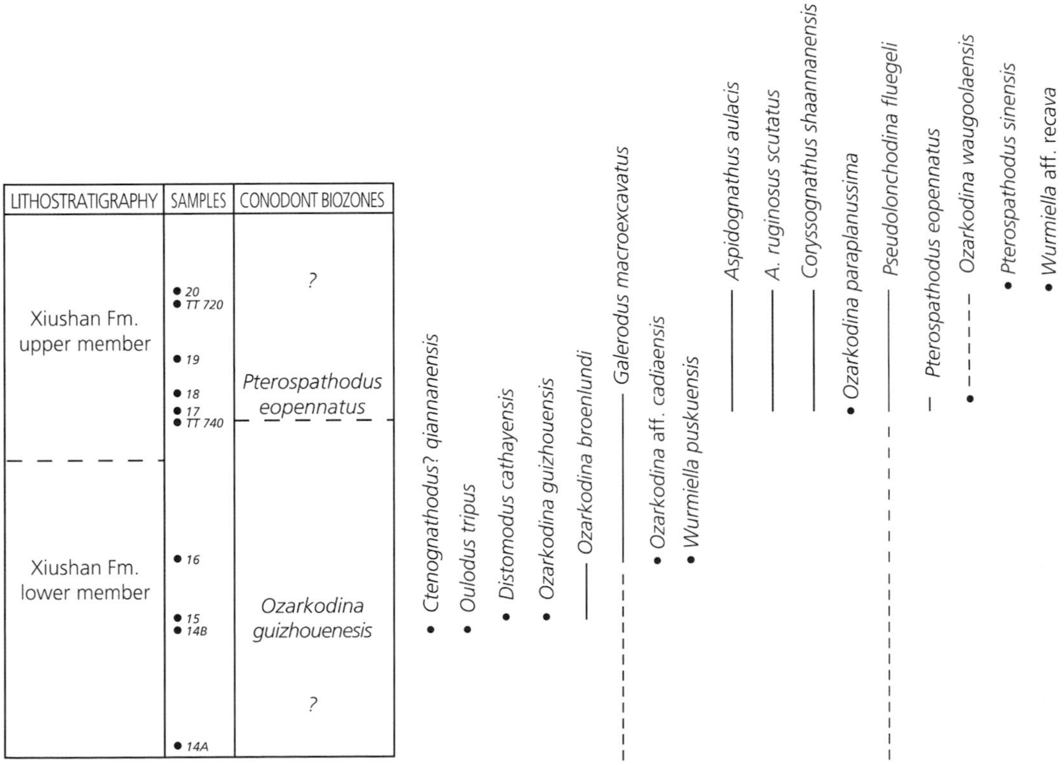

TEXT-FIG. 5. Range chart of biostratigraphically important conodonts from the Xiushan Formation, Leijiatun section.

with occasional fine-grained sandstone and marl beds. Rare brachiopods and bivalves occur, and the presence of the chitinozoan *Eisenackitina daozhenensis* (Geng) in the lower part of the formation suggests an early Telychian age (Geng 1990; Chen and Rong 1996). Samples from the shales have not yielded conodonts.

The Xiushan Formation, 450 m thick, is divided into lower and upper members. The lower member comprises yellow-green shales with fine-grained sandstones and siltstones. Macrofossils are uncommon and of very low diversity, but brachiopods, bivalves and trilobites have been collected from the shales. Rare lenses of coarse biosparite occur, and these are rich in fossil remains, including trilobite fragments, gastropods, ostracodes and conodonts.

The upper member is separated from the lower member by a covered interval and is very rich in macrofossils. It consists of shales and mudstones, sometimes silty, with thin beds and lenses of biosparitic limestone, which are variably dolomitized. Graptolites, chitinozoans and conodonts all indicate a late Telychian age (Chen and Rong 1996).

The uppermost 39 m of Silurian strata are assigned to the Huixingshao Formation, formerly the 'upper marine red beds'. The formation is unconformably overlain by Permian sandstone, and consists of purple-red sandy shales and mudstones. No conodonts have been recovered from the Huixingshao Formation, but scarce bivalves, corals and eurypterid fragments are recorded (Chen and Rong 1996). The precise age is uncertain.

The Yushitan section

The Ningqiang Formation and the underlying Wangjiawan Formation are well developed and exposed in a nearly continuous section, the Yushitan section (sometimes also referred to as the Yushitan-Xiaoshizuizigou section) near Ningqiang county town, southern Shaanxi Province (see Chen *et al.* 2002, pp. 19–20; Text-fig. 6). This section is briefly summarized below and a summary measured section shown in Text-figure 7.

The Wangjiawan Formation sits conformably on the graptolitic shales and siltstones of the Cuijiagou Formation, which was not sampled in this study. The Wangjiawan Formation is 344 m thick, dominantly comprising purple and grey-yellowish silty shales and siltstones, which have yielded graptolites, chitinozoans and other fossils indicating a possible age range from the *crispus* to *griestoniensis* graptolite biozones (Chen and Rong 1996; Chen *et al.* 2002). Nine metres of limestone occur at the base of the formation, and samples from this have yielded conodonts (including sample Ningqiang 7 from the top of the thin-bedded, stylolitic lower half; Ningqiang 6 from the massive upper bed, 1.5 m below the top of the limestone).

The Ningqiang Formation is divided into two members, the 1232-m-thick Yangpowan Member, and the Shenxuanyi Member, seen to 347 m in this section. The Yangpowan Member is characterized by grey-green shales, with intercalations of nodular and bioclastic limestones and reddish shales. Some small bioherms or slumped fossiliferous limestone blocks are developed in the middle part of the member.

The Shenxuanyi Member consists of grey-yellow to grey-green shales with thin bioclastic limestone beds and bioherms. Sample Ningqiang 1 is from 3 m below the base of a prominent bioherm, the Yushitan bioherm (Qiu J. Y. 1990), and sample Ningqiang 2 is from 3 m above a higher bioherm, near the top of the undisturbed exposure.

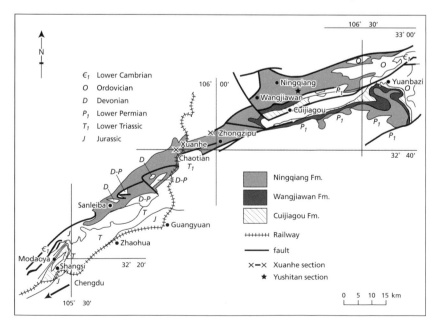

TEXT-FIG. 6. Geological map of the Guangyuan-Ningqiang region showing the locations of the Yushitan and Xuanhe sections (modified after Chen *et al.* 2002).

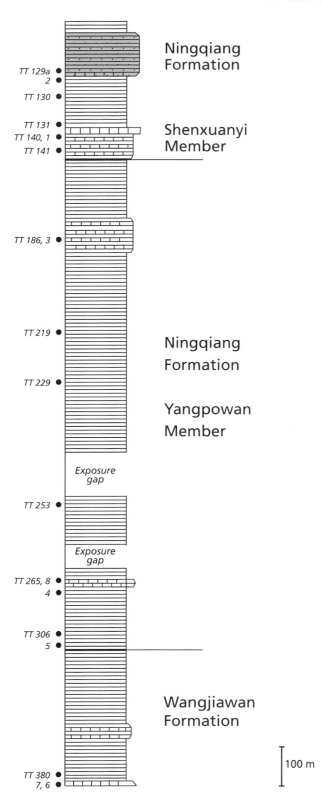

Ningqiang
Formation

TT 129a ●
2 ●

TT 130 ●

TT 131 ●
TT 140, 1 ●
TT 141 ●

Shenxuanyi
Member

TT 186, 3 ●

TT 219 ●

TT 229 ●

Ningqiang
Formation

Yangpowan
Member

*Exposure
gap*

TT 253 ●

*Exposure
gap*

TT 265, 8 ●
4 ●

TT 306 ●
5 ●

Wangjiawan
Formation

100 m

TT 380 ●
7, 6 ●

TEXT-FIG. 7. Summary measured section for the Yushitan section, showing the sample horizons of samples Ningqiang 1–8 (numbers only shown) and of key TT samples.

A range chart of key conodont taxa is shown in Text-figure 8, and full data on conodont occurrences are given in Tables 6–7 (see Appendix).

The Xuanhe section

Xuanhe (Shenxuanyi) is a small town in north-west Sichuan Province, located between Guangyuan city and Ningqiang county town (Text-fig. 6). The composite sampled section, exposed up a steep hillside, is regarded as representative of widely distributed Telychian rocks in the area (Chen *et al.* 2002, p. 14); it is the type locality of the Shenxuanyi Member (upper part of the Ningqiang Formation). The Shenxuanyi Member contains a latest Telychian fauna, with graptolites of the *spiralis-grandis* Biozone occurring in the lower part (Chen and Rong 1996); it is unconformably overlain by the Devonian (Givetian) Lungdongbei Formation. The Wangjiawan Formation conformably underlies the Shenxuanyi Member, but is not well exposed in the Xuanhe section.

There is uncertainty about the true thickness of the Shenxuanyi Member as there appears to be some repetition of beds; Jin *et al.* (1989) attributed this to an overturned syncline, but it may be that faulting rather than folding is responsible. The directly measured total thickness is about 1783 m (Chen and Rong 1996), but Jin *et al.* (1992) gave a figure of only about 847 m. Four major limestone units (L1–L4) are recognized in the composite section (Text-fig. 9) collected during the present study (see Chen and Rong 1996), separated by purple shales and siltstones; above the top limestone are grey-green shales with interbedded nodular and bioclastic limestones. Samples Xuanhe 1 and 2 are from the base and top of L1, Xuanhe 3 and 4 from the base and top of L2, Xuanhe 5 and 6 from the base and top of L3, and Xuanhe 7 and 9 from the base and top of L4; Xuanhe 8, 10 and 11 are from above the fourth limestone; other samples, prefixed TT, were taken throughout the member. The ranges of conodonts found in the Xuanhe section are shown in Text-figure 10 and details of occurrences in Tables 8–9 (see Appendix).

The Huanggexi section

Silurian strata are well developed and exposed in the area around Huanggexi, a very small village, 15 km north of Daguan county town in north-eastern Yunnan. The Huanggexi section is located to the north of the village near the road from Daguan County, Yunnan, to Yanjun County, Sichuan.

This section was first studied by Guo and Huang (1942). They classified the Silurian strata of the Daguan

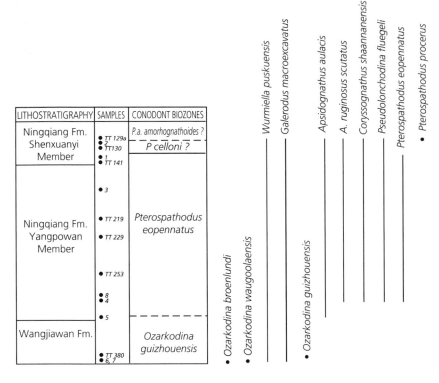

TEXT-FIG. 8. Range chart of biostratigraphically important conodonts from the Yushitan section.

area into a Lower Silurian Lungmachi shale, a Middle Silurian Daguan Group and an Upper Silurian red shale. Subsequently, Lan Chao-hua *et al.* worked in the area in 1975 and 1976 and considered that the Huanggexi section was the best available in east Yunnan, representing nearly the entire Silurian sequence. Lan (1979) named three new formations equivalent to the Daguan Group of Guo and Huang (1942): the Huanggexi, Sifengya and Daluzhai formations. Lan's classification of the Silurian sequence is as follows:

Upper Silurian: Caidiwan Formation
Middle Silurian: Daluzhai Formation
         Sifengya Formation
Lower Silurian: Huanggexi Formation
         Lungmachi Formation

Lin (1979) recommended this section as a parastratotype section representing the Middle Silurian strata of south China; he used the terms 'Baisha Stage' and 'Xiushan Stage' for the Middle Silurian sediments. However, the presence of Middle Silurian deposits remains to be demonstrated. In June 1990, a research group from the Nanjing Institute of Geology and Palaeontology investigated this section, and Wang Cheng-yuan collected conodont samples. These produced Llandovery conodonts, and there is no evidence for Middle Silurian marine deposits. The Nanjing team also re-identified supposed Wenlock graptolites as Llandovery forms and re-designated the Sifengya and Daluzhai as members of

the Takuan Formation (Chen and Rong 1996; Chen *et al.* 2002).

The base of the section is taken at the Kuanyinchao Bed, here consisting of 1.7 m of black shales with limestone lenses. Hirnantian brachiopods occur together with trilobites, and a few conodonts, including *Sagittodontina* sp., have been recovered; the fauna indicates a latest Ordovician age.

The overlying Lungmachi Formation comprises dark grey, thin-bedded argillaceous limestones, intercalated with black shales and sandstones. The top of the formation is considered to be no younger than the *triangulatus* Graptolite Biozone (Chen and Rong 1996; Chen *et al.* 2002). No conodonts have been recovered.

The Huanggexi Formation is 155 m thick and dominated by massive nodular and argillaceous limestones, intercalated with yellow-green shales and thin-bedded limestones. It contains a diverse macrofauna, including trilobites, brachiopods, corals, gastropods and some graptolites, and correlates partly with the *turriculatus* Biozone (Chen and Rong 1996; Chen *et al.* 2002). Conodonts (samples TT 1142–1138) include *Ozarkodina obesa*.

The Sifengya Member is 167 m of purple to red shales intercalated with thin-bedded argillaceous limestones, yielding brachiopods, corals, trilobites and some conodonts. Chen and Rong (1996) suggested that the Sifengya Member correlates with the Rongxi and Majiachong formations. The succeeding 359 m of the Daluzhai

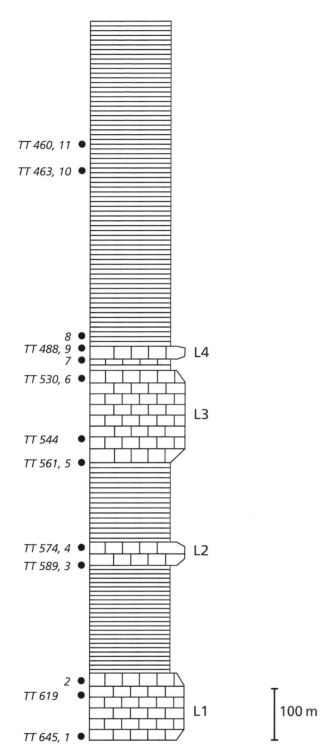

brachiopods, trilobites, bivalves, corals, nautiloids and graptolites. Conodonts (TT 1153–1159, TT 1164–1169) include species of *Apsidognathus*, *Aulacognathus* and *Pterospathodus*, indicative of a Telychian age and consistent with a correlation with the upper member of the Xiushan Formation in Guizhou and the Ningqiang Formation in Shaanxi, as suggested by Chen and Rong (1996).

The Caidiwan Formation, 211 m thick and overlain unconformably by Devonian strata, is dominated by purplish-red shales and sandstones and is considered to be equivalent to the Huixingshao Formation (Chen and Rong 1996; Chen *et al.* 2002). No conodonts have been recorded.

The Erlangshan section

The Erlangshan area is situated 160 km south-west of Chengdu, in western Sichuan Province; the eastern part is in Tianquan County and the western part in Luding County. Tectonically, the region forms part of the western margin of the Yangtze Platform, flanking the south-eastern margin of the Qinghai-Xizang Plateau. A research team led by Jin Chun-tai investigated the area between 1983 and 1985, reporting their results in a monograph (Jin *et al.* 1989); conodont discoveries were recorded using form taxonomy by Yu Hong-jin. These authors named a sequence of formations in the area, represented in eight measured stratigraphical sections, of which they considered the Yuanyangyan-Maliuqiao section to be the best (see Chen *et al.* 2002, fig. 15; Text-fig. 11). New conodont samples reported in the present paper are all from this section. The stratigraphic sequence is briefly described below, using the traditional three-fold division for the Silurian adopted by Jin *et al.* (1989).

The 'Lower Silurian' of Jin *et al.* (1989) comprises two formations: the Yuanyangyan and Luoquanwan formations. The Yuanyangyan Formation, 256 m thick, rests conformably on the Upper Ordovician Erlangshan Formation and comprises black mudstones, black shales and sandstones with a rich graptolite fauna that includes *Parakidograptus acuminatus* and *Monograptus sedgwickii*; no conodonts have been recovered. The Luoquanwan Formation is about 500 m thick and consists of calcareous mudstones, sandstones and black shales bearing graptolites; again, no conodonts have been recovered.

The 'Middle Silurian' of Jin *et al.* (1989) consists of three formations, all of which are now considered to be of Telychian age on the basis of rich macrofaunal and conodont collections (Chen *et al.* 2002). The Longdanyan Formation comprises 88 m of argillaceous and nodular limestones. The list and figures of conodonts presented by Yu (*in* Jin *et al.* 1989) include several elements of

TEXT-FIG. 9. Summary measured section for the Xuanhe section, showing the four major limestone developments (L1–L4) and sample horizons of samples Xuanhe 1–11 (numbers only shown) and of key TT samples.

Member comprise argillaceous limestones, intercalated with calcareous siltstones, calcareous sandstones and black shales. There is a rich and diverse fossil fauna, with

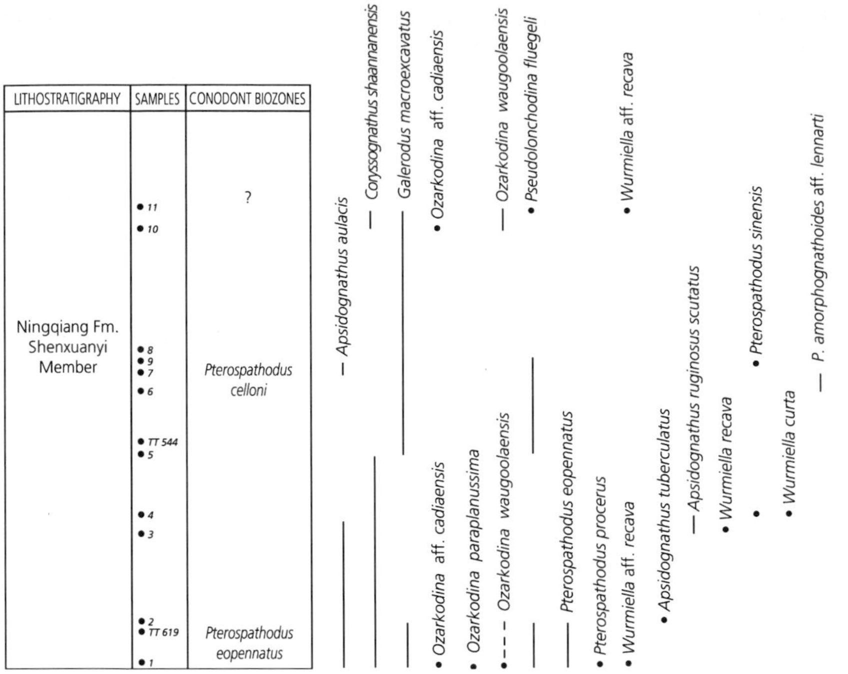

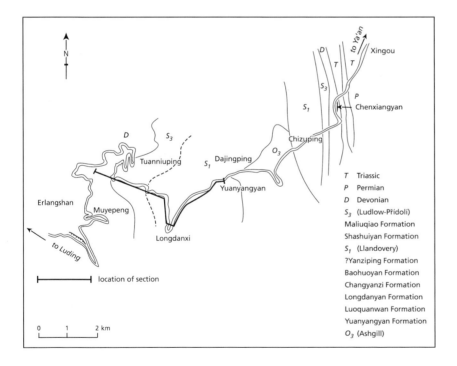

TEXT-FIG. 10. Range chart of biostratigraphically important conodonts from the Xuanhe section.

TEXT-FIG. 11. Outline geological map showing the Yuanyangyan-Maliuqiao section, Erlangshan, south-west Sichuan Province (modified after Chen *et al.* 2002).

Distomodus, possibly referable to *D. staurognathoides*, unequivocally indicating a Llandovery age. The overlying Changyanzi formation is 439 m thick, predominantly of grey calcareous and silty mudstones, with limestone lenses. Again, the conodonts recorded by Yu (*in* Jin *et al.* 1989) demonstrate a Llandovery (Telychian) age, particularly the presence of *Aulacognathus bullatus*. The Baohuo-yan Formation is 523 m thick and divided into three members, respectively 399, 112 and 12 m in thickness. The lower member is dominated by limestones, the middle member is dolomitic, and the upper member comprises thick-bedded crystalline limestones. Conodonts include *Ozarkodina* cf. *parahassi* and *Pseudolonchodina fluegeli*, indicative of a Llandovery age (Chen *et al.* 2002).

The 'Upper Silurian' of Jin *et al.* (1989) also includes three formations. The lowest, the Yanziping Formation, consists of 279 m of purple-red mudstones overlain by grey-green dolomites and sandstones; no macrofossils or conodonts have been recorded, and the age of this formation is equivocal. The overlying Sashuiyan Formation, comprising 177 m of limestones and mudstones, has yielded conodonts, including *Ozarkodina crispa*, indicative of a latest Ludlow age (Yu *in* Jin *et al.* 1989). *Ozarkodina crispa* has also been recovered in the present study from the lower part of the overlying Maliuqiao Formation indicating that this part of the formation, at least, is also of Ludlow age. The Maliuqiao Formation comprises 117 m of limestones and mudstones followed by grey dolomites.

MATERIAL AND METHODS

Samples collected by WCY are prefixed 'TT' and were processed and picked in Nanjing. Those collected by RJA are prefixed with a locality indicator: 'Shiqian' for the Leijiatun section, 'Ningqiang' for the Yushitan section, and 'Xuanhe'. In total, more than 20 000 conodont elements have been recovered and studied from the sampled sections. The repositories for figured specimens are the Nanjing Institute of Geology and Palaeontology, Academia Sinica (NIGPAS) and the Natural History Museum, London (NHM). All 536 specimens newly illustrated here are housed in NIGPAS with catalogue numbers 149623 to 150158; four specimens are re-illustrated, two from the NIGPAS collections (Pl. 10, figs 1–2, 117112; Pl. 23, fig. 10, 117115) and two from the NHM (Pl. 18, fig. 21, NHM X1133; Pl. 25, figs 13–14, NHM X1142).

Most samples were of limestone and were processed using buffered acetic acid (see Jeppsson *et al.* 1985). Residues were concentrated using bromoform or sodium polytungstate as heavy liquids. Some residues were still large after this treatment and were further concentrated using magnetic separation or di-iodomethane. Even then, some residues with a large content of phosphatic/phosphatized material required many hours of hand-picking. Shale samples were disaggregated using hydrogen peroxide and/or petroleum ether, but none yielded conodont elements. These techniques are all described in the first three chapters of Austin (1987).

SILURIAN CONODONT ZONATION IN CHINA

The first conodont zonation for the Silurian was proposed by Walliser (1964), whose scheme was based on the section at Mount Cellon in the Carnic Alps of Austria. Walliser's lowest unit, Bereich 1, has not been adopted by subsequent workers as its age is uncertain and it is overlain by a hardground representing a considerable hiatus (Schönlaub 1971). Above this gap, Walliser named ten Silurian conodont zones, beginning with the *celloni* Zone and ranging to the *eosteinhornensis* Zone; the lowermost part of the succeeding *woschmidti* Zone may also be within the Silurian (see Aldridge and Schönlaub 1989). It is possible that the lowermost part of the *celloni* Zone is not represented at Cellon, although the lowest specimens of *Pterospathodus celloni* (Walliser) compare with those known from the base of the zone elsewhere (Männik and Aldridge 1989). It has also become clear that there are several gaps in the condensed sequence exposed at Cellon. The gap below the *celloni* Zone is probably the most significant of these as it encompasses all or most of the pre-Telychian Silurian and probably some of the Telychian. After 1964, a number of local conodont biozonal schemes were proposed for this lower Silurian interval. These were reviewed by Aldridge and Schönlaub (1989; see Text-fig. 12), who suggested that the lowest Silurian strata should be accommodated in two standard biozones, named after *Distomodus kentuckyensis* and *D. staurognathoides*.

Since 1989, there has been continuing establishment and refinement of local and global conodont biozonations for parts of the Silurian. Of particular relevance to the present study are two proposed global biozonations, published by Zhang and Barnes (2002) and Männik (2007), that together accommodate almost the entire Llandovery Series (Text-fig. 12). In addition, Jeppsson (1997) established a detailed biozonation that spans the uppermost Telychian and lower Wenlock and includes subdivisions of the *Pterospathodus amorphognathoides amorphognathoides* Zone.

Zhang and Barnes (2002) noted that *Distomodus* exhibits a preference for high-energy inshore environments and suggested that a more generally applicable zonation for the earliest Silurian should be based on taxa found in more widely distributed open shelf environments, principally species of *Ozarkodina* and, to a lesser extent, *Oulodus*. They, therefore, proposed a new zonation founded on the succession they studied on Anticosti Island, Québec (Text-fig. 12): the *Ozarkodina hassi* Zone, the *Ozarkodina strena* Zone, the *Ozarkodina clavula* Zone and the *Ozarkodina aldridgei* Zone. The base of each of these zones is defined at the first appearance of the nominate species. Of these taxa, *O. hassi* (Pollock *et al.*) has been reported from many parts of the world, *O. aldridgei* Uyeno is known from Europe and North America, *O. clavula* Uyeno is known from Anticosti Island and Severnaya Zemlya, and *O. strena* Zhang and Barnes has, to date, only been reported from Anticosti Island. The *O. strena* Zone was subdivided into two subzones, based on the first appearances of *Oulodus*

SERIES AND STAGES		'GLOBAL STANDARD'	ANTICOSTI ISLAND	BALTIC AREA		CHINA
		ZONES	ZONES	SUPERZONES	ZONES	ZONES
Wenlock (base of)		Pterospathodus amorphognathoides		Pseudooneotodus bicornis	Upper Ps. bicornis	
					Lower Ps. bicornis	
Llandovery	Telychian				P. a. amorphognathoides	P. a. amorphognathoides
		Pterospathodus celloni		Pterospathodus celloni	P. a. lithuanicus	Pterospathodus celloni
					P. a. lennarti	
					P. a. angulatus	
		Distomodus staurognathoides		Pterospathodus eopennatus	P. eopennatus ssp. 2	Pterospathodus eopennatus
					P. eopennatus ssp. 1	
	Aeronian		Ozarkodina aldridgei			Ozarkodina guizhouensis
			Ozarkodina clavula			- - - - - - - - - unzoned interval
		Distomodus kentuckyensis	Ozarkodina strena / Oulodus jeannae			- - - - - - - - -
	Rhuddanian		Ozarkodina strena / Oulodus panuarensis			Ozarkodina obesa
						- - - - - - - - -
			Ozarkodina hassi			Ozarkodina aff. hassi

TEXT-FIG. 12. Conodont biozonations for the Llandovery Series. 'Global Standard' biozonation after Aldridge and Schönlaub (1989). Anticosti Island biozonation after Zhang and Barnes (2002), who also illustrated the relationship of this scheme to previous local biozonations. Baltic biozonation after Männik (2007) and, for the uppermost Telychian and lower Wenlock, Jeppsson (1997). China biozonation based on the present study. The precise stratigraphic relationship between the top of the Ozarkodina aldridgei Zone and the base of the Pterospathodus eopennatus Zone is unknown.

panuarensis Bischoff and Oulodus jeannae Schönlaub. The top of the O. aldridgei Zone is taken at the first appearance of Pterospathodus celloni.

Männik (2007), following extensive studies of conodonts from Estonia, proposed a major revision of biozonation in the interval characterized by forms assigned to Pterospathodus celloni (Text-fig. 12). He noted that many specimens assigned to this species in the literature, including those from China, did not conform closely to the type material and erected a new species Pterospathodus eopennatus, which accommodates most of these. He further proposed two superzones: the P. eopennatus and P. celloni superzones. The P. eopennatus Superzone is divided into the P. eopennatus ssp. n. 1 and P. eopennatus ssp. n. 2 zones, and the P. celloni Superzone is divided into the P. amorphognathoides angulatus, P. a. lennarti and P. a. lithuanicus zones. Each zone is regarded as corresponding to the total range of the nominate taxon, with the boundaries between the zones recognized by the evolutionary replacement of one taxon in the P. eopennatus – P. amorphognathoides lineage by another. The top of the P. a. lithuanicus zone occurs at the first appearance of P. a. amorphognathoides.

In China, Llandovery marine deposits are mostly represented in the upper and lower reaches of the Yangtze Valley, whereas Wenlock to Přídolí marine deposits are mainly distributed in west Yunnan, Tibet, Xinjiang and Inner Mongolia. The conodont biostratigraphical sequence has been summarized several times (Wang C. Y. and Wang Z. H. 1981, 1983; Lin 1986; Wang C. Y. 1990; Wang and Aldridge 1996; Aldridge and Wang 2002). Conodont faunas representing much of the Silurian have been recognized in China, but there is sparse knowledge regarding the Wenlock and Přídolí. The present paper deals primarily with Llandovery conodonts from south China, together with some information on Ludlow conodonts from the Erlangshan area. By combining this new information with published data, we have been able to refine the local conodont biozonation for the lower and middle parts of the Llandovery and to assess the local relevance of the schemes proposed by Zhang and Barnes (2002) and Männik (2007) (Text-fig. 12). Current knowledge of the Silurian conodont biostratigraphical sequence in China is summarized below in ascending order of biozones; all the biozones are successive appearance biozones, with the base of each defined by the first appearance of the eponymous species.

Llandovery biozones

Ozarkodina aff. *hassi* Biozone

The lowest zone named by Zhou *et al.* (1985), the *Panderodus gracilis* – *P. simplex* Assemblage Zone, is based on very long-ranging form species and is of minimal biostratigraphic value. However, we have found specimens closely similar to *Ozarkodina hassi* in the Kuanyinqiao Bed (TT 840, Shiqian-1, Shiqian-2) at Leijiatun. On Anticosti Island, the first occurrence of *O. hassi* was taken by McCracken and Barnes (1981, p. 72) to be indicative of the base of the Silurian and was used by Zhang and Barnes (2002) to define the base of the *O. hassi* Zone. *O. hassi* has also been documented from Aeronian and lower Telychian strata of the Welsh Borderland (Aldridge 1972, 1975) so it has a long range. However, the first appearance of conodonts of this type is very close to the Ordovician/Silurian boundary, and we propose provisionally to use *O.* aff. *hassi* as a local zonal fossil for the uppermost Ordovician – lowest Silurian interval below the *O. obesa* Biozone.

Ozarkodina obesa Biozone

This zone was named by Zhou *et al.* (1981). They found the eponymous species in the middle and upper parts of the Xiangshuyuan Formation at Leijiatun, where our collections extend the range into the lower part of the formation (samples Shiqian 4, 5). We have also found *O. obesa* in the Leijiatun Formation, although the related new species *O. wangzhunia* is much more common. Zhou *et al.* (1981) found few conodonts in the Leijiatun Formation and assigned it to their 'Interval-Zone I (Zone A)', but we extend the *O. obesa* Biozone into this formation. The presence of *O. pirata* in the middle and upper parts of the Leijiatun Formation (samples Shiqian 8, 9) provides a potentially useful additional stratigraphic marker, although this species has a very long range in Anticosti Island, from the base of the *O. strena* Zone to the lower *O. aldridgei* Zone (Zhang and Barnes 2002, fig. 8). The *O. obesa* Biozone is equivalent to the upper part of the *D. kentuckyensis* Biozone of the standard zonation and may extend into the lower *D. staurognathoides* Zone.

Ozarkodina parahassi Biozone

Zhou *et al.* (1981) proposed a '*Spathognathodus parahassi* – *Spathognathodus guizhouensis* Assemblage Zone' below the *celloni* Zone, based on their collections from Leijiatun. They divided this assemblage zone into two subzones, a lower *S. parahassi* Subzone and an upper *S. guizhouensis* Subzone, although they did not formally define the bases of these units. The *S. parahassi* Assemblage Subzone was not based on the Leijiatun section, but on the occurrence of the species in strata at Tudiao, Yanhe County, considered by Zhou *et al.* (1981) to be equivalent to the Rongxi Formation. Ding and Li (1985) also identified (but did not illustrate) '*S.*' *parahassi* in the Ningqiang area, but did not record '*S.*' *guizhouensis*, so they gave the lowest zone they recognized the name '*Spathognathodus parahassi* Assemblage Zone'.

We have only recovered a few specimens of *O. parahassi* from the Leijiatun section, and these occur within the range of *O. obesa*. We cannot, therefore, confirm the value of this zone. One sample that has yielded *O. parahassi* (Shiqian 9) also contains *O. pirata* and specimens referred herein to *O.* cf. *parainclinata*; these latter specimens might represent an early form of *O. guizhouensis* (see below). These specimens formed part of the basis used by Wang and Aldridge (1996) for extension of *O. guizhouensis* downwards to encompass the range of *O. parahassi*.

Zonation of this interval is complicated by the paucity of conodonts from the Majiaochong Formation (Aldridge and Wang 2002, p. 91) and their absence from samples from the Rongxi Formation and the lower part of the lower member of the Xiushan Formation at Leijiatun. We, therefore, leave this interval unzoned in respect of conodonts. The occurrence of *Ctenognathodus? qiannanensis* below the first appearance of *O. guizhouensis* may be stratigraphically helpful, but this appears to be a species characteristic of a restricted, nearshore facies.

Ozarkodina guizhouensis Biozone

Ozarkodina guizhouensis is widely distributed in south China, and an *O. guizhouensis* Biozone was formalized as part of the Transhemisphere Telychian project (Chen and Rong 1996; Holland and Bassett 2002), with its base at the first appearance of the eponymous species. The *O. guizhouensis* Biozone equates to part of the *D. staurognathoides* Biozone, erected by Aldridge (1972) in Britain and ranging from the Aeronian into the Telychian. In China, *D. staurognathoides* has been found in north Tibet (Yu 1985), with rare specimens from south China also identified as this species.

Pterospathodus eopennatus Biozone

This zone was introduced by Männik (1998), who subsequently (Männik 2007) raised it to superzone status, encompassing two zones that are each characterized by an unnamed subspecies of *P. eopennatus* (see above). In the Leijiatun section, *P. eopennatus* first appears in the lowest samples from the upper member of the Xiushan Formation. As noted in the systematic section below, we are unable to refer the populations of *P. eopennatus* we have studied unequivocally to the subspecies recognized by Männik (1998). It should be noted that the two sequential unnamed new subspecies of Männik (1998) contain several identical morphotypes and may be impossible to differentiate in small collections.

Pterospathodus celloni Biozone

The base of this zone (or superzone of Männik 2007) is defined at the first appearance of *P. celloni* (Walliser 1964). Following the re-assessment by Männik (1998), it is uncertain whether any true specimens of *P. celloni* have been found in China, although the species has been very widely reported, for example by Zhou *et al.* (1981), from the Xiushan Formation at Leijiatun and from the Lomiang Formation at Kaili, Guizhou Province. Zhou *et al.* (1985) also reported *P. celloni* from Sichuan and Shaanxi. Further records of this zone were provided by Ding and Li (1985) from the Ningqiang area, by Yu (1985) and Lin and Qiu (1983, 1985) from Tibet, and by Zuo (1987) and Wang G. X. *et al.* (1988) from W Hunan. All of these records need to be reviewed. Specimens of *P. celloni* reported by Qiu H. R. (1985, 1988), by Jiang *et al.* (1986) and by Wang and Aldridge (1996) were regarded as synonymous or equivocally synonymous with *P. eopennatus* by Männik (1998).

The almost certain presence of the *P. celloni* Biozone in the Xuanhe section is indicated by the records of *P. amorphognathoides* aff. *lennarti* in samples Xuanhe 6 and TT 498, above the highest occurrences of *P. eopennatus* in sample Xuanhe 2. *P. a. lennarti* was used by Männik (2007) to define the middle of three zones within the *P. celloni* Superzone. The presence of the lowest of these three zones, characterized by *P. amorphognathoides angulatus*, is attested to by the record of the nominate taxon (as *Spathognathodus pennatus angulatus*) in the Ningqiang Formation by Qian (*in* Jin *et al.* 1992, p. 62, pl. 3, fig. 5). Another possible indicator of the *P. celloni* Biozone in our collections is *Pterospathodus sinensis*, which occurs above the last appearance of *P. eopennatus* in the Leijiatun and Xuanhe sections, although some equivocal specimens co-occur with *P. eopennatus* in sample Xuanhe 1.

Pterospathodus amorphognathoides amorphognathoides Biozone

The lower boundary of this zone was defined by Walliser (1964) at the first appearance of *P. amorphognathoides* (now *P. a. amorphognathoides*); the zone was recorded as spanning the Llandovery/Wenlock boundary in the stratotype section at Leasows, Shropshire, by Mabillard and Aldridge (1985). However, the more refined zonation proposed by Jeppsson (1997) restricts the *P. amorphognathoides* Biozone to the lower part of the range of the eponymous taxon, succeeded by the Lower and Upper *Pseudooneotodus bicornis* biozones, which also contain *P. amorphognathoides*. In this formulation, the *P. amorphognathoides* Biozone is restricted to the uppermost Llandovery. Männik (2007) further subdivided the *P. amorphognathoides amorphognathoides* Zone into lower and upper subzones, with the base of the upper subzone defined by the first appearance of his *Aspelundia fluegeli* ssp. n. (= *Pseudolonchodina fluegeli* ssp. n.).

P. amorphognathoides was first reported in China by Wang C. Y. and Wang Z. H. (1981, p. 107) and by Ni *et al.* (1982) from the Renheqiao Formation of Shidian County, western Yunnan. It has also been recorded in the Pulu Formation (Wang and Ziegler 1983) and the Kede Formation (Lin and Qiu 1983; Qiu H. R. 1985, 1988), both in Dingri County, south Tibet, as well as in the Zhanongema Group of the Xainza area, north Tibet (Yu 1985). Xia (1993) reported the lower *amorphognathoides* Zone at Hanjiga Mountain, north Xinjiang, and the zone has also been recognized in Yanbian, Sichuan (An 1987; He and Qian 2000; Jin *et al.* 2005; Wang C. Y. *et al.* 2009). However, throughout much of south China (Hunan, Guangxi, Guizhou and elsewhere) the zonal fossil has not been found, and we have not recovered it from the samples from the Yangtze Platform collected for the present study.

Wenlock biozones

Walliser (1964) identified two post-*amorphognathoides* biozones in the Wenlock, named after *Kockelella patula* Walliser and '*Spathognathodus*' *sagitta* Walliser. As noted by Aldridge and Schönlaub (1989), *K. patula* is uncommon and has not been widely enough reported to be a useful international index species; also, recognition of the *O. sagitta* Zone *sensu* Walliser (1964) should be based only on the nominate subspecies, as this is the only subspecies to occur in Walliser's reference section at Cellon. Aldridge and Schönlaub (1989, p. 278), therefore, suggested the use of three standard biozones for this interval: the *Ozarkodina sagitta rhenana* Biozone, the *O. s. sagitta* Biozone and the *O. bohemica bohemica* Biozone. All are successive appearance biozones. A much more detailed zonation spanning most of the Wenlock Series was proposed by Jeppsson (1997), but this has yet to be applied anywhere in China. Wang C. Y. (1990) used the *Kockella ranuliformis* Zone as an equivalent to the *K. patula* Zone, but the two do not equate; in Jeppsson's scheme the Lower and Upper *K. ranuliformis* zones are separated from the *K. patula* Zone above by three intervening biozones.

Ozarkodina s. sagitta s. s. was mentioned as occurring in China by Lin (1986), but there is no illustration or description to support this report. *Ozarkodina b. bohemica* was recorded by Lin and Qiu (1983) and by Qiu H. R. (1985) from the Kede formation in Tingri County, south Tibet. It has also been reported from the Zhanongema Group in the Xainza area, north Tibet (Yu 1985), and from Diepo County, Gansu Province (An 1987). Wenlock conodonts are widely distributed in Tibet and west Yunnan, but there are no reliable records from the Yangtze platform.

Ludlow biozones

Walliser (1964) proposed six Silurian conodont zones spanning the Ludlow and Prídolí series: the *crassa*, *ploeckensis*, *siluricus*, *latialatus*, *crispus* and *eosteinhornensis* zones. The *crassa* Zone is based on a poorly known Pb element, and Aldridge and Schönlaub (1989) recommended that this zone should not be used, instead placing the *Ancoradella ploeckensis* Biozone directly above the *Ozarkodina bohemica bohemica* Biozone. They also suggested that *Ozarkodina snajdri* (Walliser) may be a more useful index species than *Pedavis latialata* (Walliser), and placed an *O. snajdri* Biozone between the *Polygnathoides siluricus* and *Ozarkodina crispa* biozones, with its base defined at the last occurrence of *P. siluricus* Walliser. A more refined sequence of conodont faunas established on Gotland by Jeppsson (1998) has yet to be widely recognized.

An additional useful indicator of an early to mid Ludlow age is *Kockelella variabilis* Walliser. Walliser (1964) recorded the range of this species to be from the *crassa* Zone to the *siluricus* Zone, while in England it has been reported from the Upper Bringewood and Lower Leintwardine formations (upper Gorstian and lowermost Ludfordian stages) (Aldridge 1975, 1985).

In China, *K. variabilis* was first recognized in the Zhongcao Formation of western Yunnan (Wang C. Y. *in* Ni *et al.* 1982). *Ancoradella ploeckensis* Walliser was first reported by Lin and Qiu (1983) and Qiu H. R. (1985) from the middle Keya Formation in the Himalaya district, south Tibet, where it co-occurs with *K. variabilis*. Lin and Qiu (1983) proposed the use of a *Kockelella variabilis* Fauna for an interval of strata, belonging to the Keya

Formation, between the *A. ploeckensis* Biozone and the *P. siluricus* Biozone. *Polygnathoides siluricus* itself occurs in the Shiqipou Formation of south Tibet (Lin and Qiu 1983) and has also been found in the Mendeeyao Formation of north Tibet (Yu 1985).

Ancoradella ploeckensis has also been found in the Bate'aobao area of Inner Mongolia (Wang P. 2004, 2005). Both *A. ploeckensis* and *Polygnathoides siluricus* are known from the Baizitian section in Yanbian County, south-west Sichuan (Jin *et al.* 2005; Wang C. Y. *et al.* 2009). As this region is regarded to be outside the Yangtze Platform (Rong *et al.* 2003), these two conodont biozones have yet to be recorded on the Yangtze Platform.

The *Ozarkodina snajdri* Biozone, identified as *O. crispa* by Qian (*in* Jin *et al.* 1992, pl. 3, figs 13, 17), occurs in the Chejiaba Formation of Guanyan, Sichuan. The *Ozarkodina crispa* Biozone has been quite widely recognized in China. The zonal fossil was first described and illustrated by Wang C. Y. (1980, 1981) from the Miaogao and Yulongshi formations of east Yunnan and it has subsequently been found in Tibet (Lin 1991; Lin and Qiu 1983, 1985; Qiu H. R. 1985), Sichuan (An 1987; Jin *et al.* 1989, 1992; Wan *et al.* 1991), Gansu (Wang C. Y. and Wang Z. H. 1981; He 1983; Li 1987), Yunnan (Wang C. Y. 1982, 2001; Fang *et al.* 1985) and Inner Mongolia (Wang P. 2004, 2005). In 1989, Walliser and Wang divided this species into four different morphotypes based on specimens from east Yunnan.

The *Ozarkodina eosteinhornensis* Biozone, as originally erected, spans the whole of the Prídolí Series and extends back into the uppermost Ludlow (Walliser 1964; Aldridge and Schönlaub 1989). The base was defined by Walliser (1964) at the first appearance of *O. remscheidensis eosteinhornensis* (Walliser), but the subspecies was regarded as too broadly conceived by Jeppsson (1975, 1988, 1989), who restricted and revised the taxon. He also named four new conodont zones for the Prídolí Series, with his *O. steinhornensis eosteinhornensis sensu stricto* Zone representing only the middle portion. In China, reports of the zone have followed the earlier, broader interpretation of this zone. The zone was first reported in China by Wang C. Y. (*in* Ni *et al.* 1982) and by Tang *et al.* (1982) from the Niushiping Formation (= *Camarocrinus* Bed) of Shidian County, west Yunnan. It also occurs in the Pazhuo Formation of the Tingri district in south Tibet (Mu and Chen 1984; Qiu H. R. 1985, 1988).

REGIONAL AND INTERNATIONAL CORRELATION

Much of the Llandovery Series is represented at the reference section at Leijiatun. The conodont biozones recognized within the succession are shown on Text-figures 4 and 5. At the base, the *Ozarkodina* aff. *hassi* Zone is represented in the Kuanyinchiao Bed and the Lungmachi Formation and may extend into the lowermost Xingshuyuan Fortmation. The first specimens of *Ozarkodina obesa* occur low in the Xingshuyuan Formation, and the *O. obesa* Zone appears to span this formation and much of the succeeding Leijiatun Formation, although *O. parahassi* occurs alongside *O. obesa* in the highest productive sample from the Leijiatun Formation. The Majiaochong and Rongxi formations have not yielded diagnostic conodonts. The lower member of the Xiushan Formation contains *Ozarkodina guizhouensis*, so at least part of the member can be assigned to the *O. guizhouensis* Zone. *Pterospathodus eopennatus* occurs in the lowest productive samples from the upper member of the Xiushan Formation, indicative of the *P. eopennatus* Zone. No zonal index species have been recovered from the upper part of the member, so the top of the *P. eopennatus* Zone cannot be identified. There is no conodont evidence of higher zones in the Leijiatun section, with the highest productive sample (TT 715) yielding only *Coryssognathus shaannanensis*, *Ozarkodina* sp. and *Panderodus* sp.

Ranges of key taxa and the conodont biozonation in the Yushitan section, Ningqiang, are shown in Text-figure 8. At least the base of the Wangjiawan Formation is assignable to the *O. guizhouensis* Zone, and therefore correlates with the lower member of the Xiushan Formation at Leijiatun. The *P. eopennatus* Zone is recognized through almost all of the Yangpowan Member of the Ningqiang Formation and extends into the lower part of the Shenxuanyi Member. A sample from high in the undisturbed part of the Shenxuanyi Member has yielded a single specimen of *Pterospathodus procerus*, suggesting a younger age for this part of the succession. Jeppsson (1997) introduced a *P. procerus* Superzone, which occurs in the early Wenlock, but this is based only on the higher part of the range of this taxon, following the extinction of *Pterospathodus amorphognathoides*. The earliest specimens of *P. procerus* have been reported just below the base of the *P. amorphognathoides* Zone (Männik 1998). There are no other post-*P. eopennatus* zonal indices from the Yushitan section.

The ranges of selected taxa in the Xuanhe section are shown in Text-figure 10. The lowest limestone unit of the Shenxuanyi Member contains *P. eopennatus*, but the level of the base of the zone is unknown. The absence of *P. eopennatus* in higher samples suggests that this limestone is high in the zone and correlates with the lower part of the Shenxuanyi Member at Ningqiang and with the upper member of the Xiushan Foramtion at Leijiatun. A complication is the record of *P. procerus* in association with *P. eopennatus* at the base of the first limestone (base of the Shenxuanyi Formation). This suggests that either *P. eopennatus* ranges higher or that *P. procerus* ranges

lower in China than elsewhere. Alternatively, the specimens assigned to *P. eopennatus* might be mis-identified, as this taxon shares some characteristics with *P. celloni* and Walliser (1964) illustrated considerable morphological variability in the latter taxon at Cellon. In any event, an anagenetic link from *P. eopennatus* to *P. procerus* may be questioned by this co-occurrence.

The presence of *P. amorphognathoides* aff. *lennarti* higher in the Xuanhe section (at the top of the third limestone) suggests that this horizon may be referable to the *P. a. lennarti* Zone of Männik (1998), which occurs below his recorded level for *P. procerus*. Thus, it seems that it is the range of *P. procerus* that is extended by the Chinese record. *Pterospathodus sinensis* occurs above the range of *P. eopennatus* at Xuanhe (in the second limestone) and at Leijiatun (sample Shiqian 20, upper member of the Xiushan Foramtion) and may, as noted above, be a useful stratigraphic marker.

There are problems with the upper beds sampled in the Xuanhe section. These contain no zonal indices, but include a number of taxa that are otherwise known only from the lowest beds of the section: *Ozarkodina* aff. *cadiaensis*, *O. waugoolaensis*, *Wurmiella* aff. *recava*, *Oulodus shiqianensis*. It may be that these are all long-ranging taxa and that their absence from the intervening strata is attributable to environmental controls, but there is no sedimentological evidence to support this. It is more likely that the repetition of beds suspected for this section (see above) is responsible for these occurrences, and that a progressive sequence of biozones cannot be recognized in the upper part of the Shenxuanyi Member here.

International correlations are complicated by the fact that several key Chinese taxa, including *O. parahassi*, *O. guizhouensis* and *P. sinensis* are unknown elsewhere. Similarly, many zonal indices used in other regions have not been found in the strata of the Yangtze Platform. The presence of *O.* aff. *hassi* in the lower sampled units of the Leijiatun section suggests a correlation with the *O. hassi* Zone of Zhang and Barnes (2002), which is widely distributed in strata of Rhuddanian age. However, the Kuanyinchiao Bed has been assigned a Late Ordovician (Hirnantian) age on the basis of its characteristic *Hirnantia* brachiopod fauna (Rong 1979), so the range of *O.* aff. *hassi* appears to extend below the base of the Silurian. *Ozarkodina obesa* (= *O. guiyangensis*) has been reported from the Vodopad Formation of the Sedov Archipelago, Russia, by Männik (2002), at a level assigned by him to the Aeronian, but has otherwise not been reported outside China. *Oulodus panuarensis*, however, is a more widely useful taxon in the lower part of the Llandovery Series. Zhang and Barnes (2002) used this species to define a lower subzone of their *Ozarkodina strena* Zone, immediately overlying the *O. hassi* Zone and spanning all but the lowest Rhuddanian. *Oulodus panuarensis* was recorded below *O. obesa* in the Volopad Formation of the Sedov Archipelago (Männik 2002, fig. 1) and was considered a characteristic taxon of the '*Distomodus pseudo-pesavis – Ozarkodina masurenensis* Assemblage Zone' of New South Wales, Australia, by Bischoff (1986). The base of this assemblage zone was equated by Bischoff (1986, fig. 10a) with the upper part of the *C. cyphus* graptolite zone, of uppermost Rhuddanian age. Specimens of *O.* aff. *panuarensis* recovered from the lowermost Xiangshuyuan Formation of Leijiatun (sample Shiqian 3), below the first appearance of *O. obesa*, would be consistent with a late Rhuddanian or early Aeronian age for this level.

The *O. guizhouensis* Zone cannot be correlated outside of China on the basis of the nominate species. However, in both the Leijiatun and Yushitan sections, the occurrence of *O. guizhouensis* coincides with that of *O. broenlundi*. In north Greenland, the lowest records of *O. broenlundi* are just below the first appearance of *Pterospathodus*, although the species ranges into strata assigned to the *P. celloni* Zone (Armstrong 1990). Männik (2007) recorded the first appearance of *Pterospathodus* specimens, in the form of *P. eopennatus* n. ssp. 1, as being low in the Telychian, so their absence from the lower member of the Xiushan Formation at Leijiatun would suggest an Aeronian or earliest Telychian age for this member. This constraint also indicates that the Leijiatun, Majiaochong and Rongxi formations are all of Aeronian age.

The widespread occurrence of the *P. eopennatus* Zone in south China allows global correlation at this level. Männik (1998) reported the zone from the lower part of the Velise Formation of Estonia and further re-attributed several previous records of *P. celloni* from many parts of the world to *P. eopennatus*. Among these correlatives are the Wych Formation of the Malvern Hills, England (Aldridge 1972), limestone megaclasts from the Early Silurian of New South Wales, Australia (Bischoff 1986) and strata in sections throughout north Greenland (Aldridge 1979; Armstrong 1990). Männik (2007) tentatively correlated the base of the *P. eopennatus* Zone with the upper part of the *Sp. turriculatus* graptolite zone, indicative of an early Telychian age. The top of the *P. eopennatus* Zone (equivalent to the top of the range of *P. eopennatus*) was provisionally placed in the lower part of the *M. crenulata* graptolite zone (Männik 2007, fig. 5).

Another taxon permitting global correlation is *Pterospathodus amorphognathoides lennarti*. Although the specimens from China differ a little from those elsewhere, they seem to represent part of a plexus of morphotypes that occurs within the middle and upper part of the *P. celloni* Zone. *Pterospathodus a. lennarti* has been reported from the middle part of the Velise Formation of Estonia (Männik 1998) and from the upper Hughley Shales of Shropshire, England (Aldridge 1972, samples Ticklerton 1–3).

Männik (2007) tentatively correlated his *P. a. lennarti* Zone to the middle part of the Telychian *O. spiralis* graptolite zone.

In the higher Silurian, the recognition of the *Ozarkodina crispa* Zone, which is well known from many parts of the world, permits wide global correlation. The zone occurs in the uppermost strata of the Ludlow Series: at Pozáry Quarry near Prague, boundary stratotype for the base of the Přídolí Series, it ranges through just over 2 m, with the highest occurrence 0.5 m below the base of the Přídolí (Kríz 1989, text-fig. 7), and at Ludlow, England, the species is only found in the very highest Ludlow strata, 0.15–0.3 m below the probable disconformity with the overlying Downton Castle Sandstone (Miller 1995; Viira and Aldridge 1998).

ARE WENLOCK MARINE DEPOSITS PRESENT ON THE YANGTZE PLATFORM?

Although the presence of Llandovery and Ludlow strata is well established in south China, there has been doubt and controversy regarding the existence of any marine deposits representing the intervening Wenlock Series. For some years, it was generally accepted that the entire 'Middle Silurian' was represented by marine deposits of the Lojoping and Shamo formations in the Yichang area of the upper Yangtze valley. However, graptolites from these strata have shown that they are mostly or entirely of late Llandovery age (Wang X. F. 1965; Ge and Yu 1974; Ge *et al.* 1977; Rong *et al.* 1984). Other candidates for a Wenlock age have included red beds of the Huixingshao Formation in the upper Yangtze valley, the Maoshan Formation in the lower Yangtze valley and the Yaojiashan Formation in eastern Yunnan. Although these red beds are representative of a paralic facies and lack age-diagnostic fossils, conodonts show that the uppermost Xiushan Formation, immediately underlying the Huixingshao Formation, is referable to the *P. eopennatus* Biozone. As currently defined, the base of the Wenlock Series falls above the base of the *P. amorphognathoides* Biozone (Mabillard and Aldridge 1985), so it seems probable that the Huixingshao Formation is at least partly of late Llandovery age and may all be pre-Wenlock.

As noted above, it has also been claimed that Wenlock marine deposits occur in the Huanggexi and Erlangshan sections. Conodonts, however, show that the uppermost Daluzhai Formation of the Huanggexi section belongs to the *P. eopennatus* or lower *P. celloni* Biozone. At Erlangshan, the presence of Wenlock strata was mainly determined from the conodont identifications presented by Yu (*in* Jin *et al.* 1989), whose zonation is discussed above. Yu claimed that of the 59 conodont form species he recorded in the 'Middle Silurian', eight provided definitive evidence of a Wenlock age, while the other 51 extended up from the 'Lower Silurian'. These eight were listed as: *Hindeodella confluens* Branson and Mehl, *H. equidentata* Rhodes, *Plectospathodus extensus* Rhodes, *P. flexuosus* Branson and Mehl, *Neoprioniodus bicurvatus* Branson and Mehl, *Lonchodina walliseri* Ziegler, *Ozarkodina media* Walliser and *O. ortuformis* Walliser; they were used to characterize a '*Hindeodella equidentata – Plectospathodus extensus* Range Zone'. Apart from *L. walliseri* and *O. ortuformis*, these names have been applied to elements of a variety of species of multielement *Ozarkodina*, including Llandovery examples. For instance, the names *H. equidentata*, *P. extensus*, *P. flexuosus* and *O. media* have been used for conodont elements from the Aeronian and Telychian of the Welsh Borderland (Aldridge 1972). In the same publication, the name *L. walliseri* was applied to specimens of Llandovery age that would now be placed in a species of multielement *Oulodus*. The specimen figured by Jin *et al.* (1989, pl. 2, fig. 15) as *Ozarkodina ortuformis* has a broad cavity and separated denticles and also probably represents an *Oulodus* species. Hence, none of the taxa highlighted by Yu are diagnostic of the Wenlock Series, and all could equally well occur in the Llandovery. In contrast, several of the taxa recorded in the '*H. equidentata – P. extensus* Range Zone' are restricted to Llandovery strata wherever else they have been recorded (see above). They provide compelling evidence for a Llandovery age for the Londanyan, Changyanzi and Baohuoyan formations in the Erlangshan section; there is no evidence for the suggestion of Yu (*in* Jin *et al.* 1989, p. 40) that species such as *A. bullatus*, *O. hassi* and '*Hadrognathus*' (= *Distomodus*) *staurognathoides* have longer ranges in Erlangshan than elsewhere. A Llandovery age is further confirmed by our collections from the Baohuoyan Formation (TT 1046–1049), which include *Apsidognathus tuberculatus* and *Pseudolonchodina* sp.

The Yanziping Formation overlies the Baohuoyan Formation at Erlangshan and represents 279 m of strata that have not yielded conodonts; its age has not been demonstrated. It is in turn overlain by the Sashuiyan Formation, which contains the late Ludlow conodont *Ozarkodina crispa*. A similar situation pertains in the Guangyuan area, Sichuan, from where the succession and faunas were documented by Jin *et al.* (1992). They reported a typical upper Llandovery fauna in the Upper Ningqiang Formation, above which the 54- to 168-m-thick Jintaiguan Formation contains no conodonts and at that time had yielded only a few *Lingula* sp. and fish scales. Subsequently, a brachiopod fauna has been recovered indicating a Ludlow age for at least part of the Jintaiguan Formation (Jin Chun-tai, pers. comm. 1996); key species are *Retziella uniplicata* (Grabau), *Protathyris nucleola* Fang, *P. xungmiaoensis* Chu and *Zoschizophoria hasta* Rong and Yang. This unit comprises red silty mudstones interbedded with grey silty mudstones and brown muddy siltstones and is overlain by the Chejiaba Formation, which contains *Ozarkodina snajdri*.

In conclusion, it is not currently possible to verify or refute the presence of marine Wenlock deposits on the Yangtze Platform, although they are, at best, much less extensive than has been claimed. Some strata above the highest conodont-bearing horizons in the Yushitan and Xuanhe sections might extend into the lower part of the Wenlock. However, confirmation of a Wenlock or other age for these rocks and for the Yanziping and Jintaiguan formations awaits the discovery of new fossil specimens. Perhaps acritarchs or chitinozoans will provide the required evidence.

Some typical Wenlock conodonts, including *Ozarkodina sagitta rhenana* (Walliser) and *O. bohemica* (Walliser), have been reported from the Baizitian section in Yanbian County, south-west Sichuan (He and Qian 2000; Jin *et al.* 2005), but these identifications are equivocal (see above). A number of Chinese geologists (Lin 1984; Cheng 1994; Li and Qian 2001; Liu *et al.* 2004; Jin *et al.* 2005) consider this area, together with Dali and Lijiang in west Yunnan, to be part of the south-west Yangtze Platform. However, Rong *et al.* (2003) concluded that these regions were not tectonically part of the Yangtze Platform. Several characteristic Wenlock and early Ludlow taxa, such as *O. s. rhenana*, *O. s. sagitta* (Walliser), *Kockelella variabilis*, *K. stauros* Barrick and Klapper, *Polygnathoides siluricus*, and *Ancoradella ploeckensis*, have been reported from west Yunnan and from Tibet (Qiu H. R. 1988; He *et al.* 2000; Jin *et al.* 2005); as these conodonts have not been recorded from the Yangtze Platform, they are not included in this study.

THE AGES OF SILURIAN RED BEDS IN SOUTH CHINA

Red beds are widely distributed in Silurian sequences of south China. Traditionally, Chinese geologists have divided them into two levels, informally termed 'the lower red beds' and 'the upper red beds' (Ge *et al.* 1977). The 'lower red beds' have been taken to include the Rongxi Formation in Guizhou and the 'Baisha Formation' in Sichuan, while the 'upper red beds', which are found over a wider area, include the Huixingshao Formation in Guizhou and Sichuan, the Guandi Formation in the Qujin area, east Yunnan, and the Caidiwan Formation in Daguan, east Yunnan. All these horizons lack age-diagnostic fossils.

The age of the Rongxi Formation is constrained by data from underlying and overlying strata. It rests on the Majiaochong Formation, which in turn overlies the Leijiatun Formation, the upper part of which contains *Ozarkodina parahassi* and *O. pirata*. The lower member of the Xiushan Formation, resting on the Rongxi Formation, also contains abundant conodonts, including *Ozarkodina guizhouensis*. The Rongxi Formation is, therefore, above the *O. obesa* Biozone and no younger than the *Ozarkodina guizhouensis* Biozone and is probably of uppermost Aeronian and/or lowest Telychian age.

The 'upper red beds' have conventionally been considered coeval, representing the late Telychian or the Wenlock. As noted above, the base of the Huixingshao Formation at Leijiatun probably falls within the *Pterospathodus eopennatus* or *P. celloni* biozone and much or all of the formation may be of Telychian age. The base of the Caidiwan Formation at the Huanggexi section similarly appears to fall within the *P. celloni* Biozone, as the uppermost part of the underlying Daluzhai Formation contains conodonts indicative of the lower part of this biozone (our samples TT 1164–1169). In the Ningqiang and Guangyuan area, the Ningqiang Formation contains the *P. eopennatus* Biozone and the overlying red mudstones of the Jintaiguan Formation here (see Jin *et al.* 1992) may be of similar age to the Huixingshao Formation at Leijiatun.

The Guandi Formation is more difficult to place. It unconformably overlies Middle Cambrian strata and contains purplish-red and dark purple silstones and silty shales alternating with marls and nodular limestones; the total thickness is more than 600 m. Fang (1979) separated the lower part of the Guandi Formation as an independent unit, which was termed the Yaojiashan Formation. This consists of yellowish-green shales and sandstones and was assigned by Fang (1979) to the Middle Silurian; the upper part of the original Guandi Formation was placed in the Upper Silurian. The Yaojiashan Formation has been correlated with the Huixingshao Formation, partly on the basis of the cephalopod *Sichuanoceras*, which has been reported from both the Xiushan Formation and the upper Guandi Formation (Pan 1987; Jin *et al.* 1992); however, this is a poorly known endemic genus and the evidence for the correlation is weak. There is also some conodont evidence (Fang *et al.* 1985); in particular, Walliser and Wang (1989) and Wang C. Y. (2001) reported *Ozarkodina crispa* from the upper Guandi Formation, and from the overlying Miaogao and Yulongsi formations, a total range in excess of 1000 m. Jin *et al.* (1992, p. 45) took this as an indication that *Ozarkodina crispa* has a very long range in China, but there is no independent way at present of testing this interpretation. Wang C. Y. (1980), however, pointed out that sedimentation in the east Yunnan area was rapid in the late Silurian, and this may account for the great thickness of the *O. crispa* Biozone. Red beds also underlie strata with *O. crispa* in other areas: the Yanziping and lower Sashuiyan formations in the Erlangshan area, and the Jintaiguan Formation in the Guangyuan-

Ningqiang area. These have all been referred to the 'upper red beds', but there is no evidence that they are of the same age as the Huixingshao Formation.

Present evidence, therefore, suggests that red beds are probably represented at three main levels in south China, rather than two: upper Aeronian to lower Telychian; upper Telychian, perhaps extending into the Wenlock; Ludlow, pre-*O. crispa* Biozone. Within these levels, it is also likely that there is some diachroneity.

FACIES CONTROL AND BIOGEOGRAPHY

During the Late Ordovician, the Yangtze Plate collided with the Cathaysian Plate, uplifting the northern and north-eastern parts of the Yangtze Plate. At the same time, a semi-restricted basin, represented by the Wufeng graptolite black shale facies, developed in the area covering the south-east upper Yangtze, the southern and eastern parts of Sichuan, and the northern part of Guizhou. In the Rhuddanian and Aeronian, the Yangtze Plate subsided, with the north and north-eastern parts becoming flooded. The transgressive area expanded in the early Telychian, with submarine rises developing on uplifted blocks (e.g. the Nanjiang-Wangcang uplift in Sichuan and the Maopinggou and Jinjiagou uplifts in Ningqiang and Shaanxi). Migration of the foreland uplift towards the north-western craton continued through the Telychian, with land forming in the northern and eastern parts of the Yangtze Plate, while a shallow marine shelf covered southern and western areas. Continuing uplift through the late Telychian finally culminated in emergence, probably by the end of the Telychian, but possibly in the early Wenlock.

A preliminary analysis of the relative frequencies of conodont taxa was undertaken by Zhou (1986), who concluded that the most diverse faunas were in offshore platform/slope and shallow marine basin facies; he did not detail distribution patterns at generic or specific level. The conodont collections from south China all come from relatively shallow-water environments and Silurian pelagic limestones have not been found. At Leijiatun, sedimentology suggests that the Rongxi and Huixingshao formations represent the shallowest environments, perhaps intertidal (Rong *et al.* 2003), although the red colouration should not be taken as an indication of nearshore deposition (Ziegler and McKerrow 1975). Fossils are rare in these formations, but Rong *et al.* (1984) suggested that both could be assigned to benthic assemblage 1. We have recovered no conodont elements from either formation. In the upper part of the Leijiatun Formation and the lower part of the Xiushan Formation, the slightly deeper (10–30 m) benthic assemblage 2 is indicated (Rong *et al.* 1984). The conodont faunas include rather robust elements, with the lower member of the Xiushan Formation dominated at different levels by *Ctenognathodus*? *qiannanensis*, *Distomodus cathayensis*, *Ozarkodina guizhouensis*, *Galerodus macroexcavatus* and *Wurmiella puskuensis*. Rong *et al.* (1984) also considered the Majiachong Formation to be referable to benthic assemblage 2, but we have found only a few fragmentary conodont specimens at this level. Rong *et al.* (1984) recognized a benthic assemblage 3 (30–60 m) macrofauna in the upper member of the Xiushan Formation; diverse conodonts occur here, but we consider this to be a reflection of oceanic state rather than purely related to depth (see Wang and Jeppsson 1994).

Palaeogeographically, several of the Llandovery taxa found in China have a cosmopolitan distribution. Examples are *Pterospathodus eopennatus*, *P. amorphognathoides*, *Pseudolonchodina fluegeli* and species of *Panderodus*, *Walliserodus*, *Pseudooneotodus* and *Decoriconus*. The Chinese fauna, however, also includes several species that have not yet been reported elsewhere, such as *Ozarkodina guizhouensis*, *O. wangzhunia*, *O. paraplanussima*, *Distomodus cathayensis*, *Apsidognathus aulacis*, *Ctenognathodus*? *qiannanensis*, *Pterospathodus sinensis* and some species of *Wurmiella*. Although some of these may be shallow-water taxa, there is no evidence that they were all restricted to nearshore environments, and their apparent geographical limitation may be attributable to other factors. Apart from *Ctenognathodus*?, all of these genera were globally distributed during this interval, but the south China area could represent a separate conodont province recognizable at species level. Several Chinese species are not widely known outside China, but some are also found in Australia. Among these are *Ozarkodina waugoolaensis* and *Galerodus macroexcavatus* (= *Pterospathodus cadiaensis* of Bischoff 1986). This indicates a palaeogeographical relationship between China and Australia. There is also a similarity to the fauna of north Greenland (Aldridge 1979; Armstrong 1990), which includes *Ozarkodina broenlundi*, *O. pirata*, *O.* aff. *hassi*, and *Pseudolonchodina* spp.

Apart from the cosmopolitan taxa, the Early Silurian conodont fauna from the Yangtze Platform is quite different from that of Britain. *Icriodella malvernensis*, *I inconstans*, *I*? *sandersi*, *Ozarkodina aldridgei*, *O. gulletensis*, *Astropentagnathus irregularis* and *Aulacognathus kuehni* are found in Britain but are unknown in the Yangtze Platform. Some of these are nearshore endemic taxa, but several have been reported from elsewhere in Europe and from North America, and *Icriodella inconstans* has been found in north Tibet (Wang C. Y. *et al.* 2004).

There is now sufficient information on Llandovery conodont faunas globally for rigorous vicariance and geodispersal analyses to be carried out, but such studies are outside the scope of this article.

SYSTEMATIC PALAEONTOLOGY

Multielement taxonomy has been employed throughout. Illustrated specimens prefixed NIGPAS are stored at the Nanjing Institute of Geology and Paleontology, Chinese Academy of Sciences (NIGPAS); in the plate descriptions all specimens, unless otherwise designated, are NIGPAS specimens. Specimens prefixed NHM are at The Natural History Museum, London. For apparatuses comprising ramiform elements or pectiniform plus ramiform elements, we have applied the PMS notation introduced by Sweet and Schönlaub (1975) and modified by Cooper (1975) and Sweet (1981, 1988). As originally devised, this scheme identifies Pa, Pb, M and Sa-Sd elements within a conodont apparatus (see Sweet, 1988, fig. 2.10); the notation does not relate to actual positions of the various types of elements within the apparatus, but expresses a locational analogy. Where positional homology can be determined, either from natural assemblages or through direct comparison with natural assemblages, then an expressly homological notation identifying P_{1-4}, M and S_{0-4} (P_n-S_n) elements can be applied (see Purnell *et al.* 2000, figs 2, 3, 5), and we have used this where appropriate. For apparatuses composed of coniform elements, a p, q, r notation was introduced by Barnes *et al.* (1979) and subsequently expanded by a number of authors, including Armstrong (1990) and Sansom *et al.* (1994). Sansom *et al.* (1994) based their scheme on the genus *Panderodus* and emphasized that they regarded it as strictly locational, only to be used where homologies between elements could be demonstrated. As the original P, M, S notation does not imply homology, but only locational analogy, we have followed a number of other authors (e.g. Barrick 1977; Sweet 1988) in applying it to coniform apparatuses where homologies of elements are uncertain.

The orientational terms 'anterior', 'posterior', 'lateral', 'inner' and 'outer' are used in the conventional sense for isolated conodont elements (see Sweet 1981, 1988), and do not refer to biological orientation in the animal. We use 'oral' for the upper, commonly denticulated surface of the element, and 'aboral' for the opening of the basal cavity. The allocation of some elements is equivocal: where a question mark precedes an element (e.g. ?Sa) this indicates uncertainty as to whether that element belongs to the apparatus being described; where it succeeds an element (e.g. Sa?), it indicates that the element is considered to belong to that apparatus, but its location is uncertain.

The classification adopted for apparatuses of coniform elements largely follows that proposed by Sweet (1988). For the apparatuses made up of 'complex' non-coniform elements, we follow the classification proposed by Donoghue *et al.* (2008), based on cladistic analyses of representative genera.

The taxa recovered in this study are listed in Tables 1–9 (see Appendix), and most are illustrated in the plates. Full systematic treatment is restricted here to new taxa and to taxa for which the Chinese material offers significant new information or allows new interpretations.

Order BELODELLIDA Sweet, 1988
Family BELODELLIDAE Khodalevich and Tschernich, 1973

Genus WALLISERODUS Serpagli, 1967

Type species. Acodus curvatus Branson and Branson, 1947, by subsequent designation of Cooper (1975, p. 995).

Remarks. Apparatuses of *Walliserodus* show some variation in our collections from south China. We refer the majority of specimens to *Walliserodus curvatus*, although the Chinese specimens differ in minor respects from the material from the type stratum, the Brassfield Formation (see Branson and Branson 1947; Rexroad 1967). In particular, the base of the M element in the Chinese material is more extended posteriorly, giving a sharper angle between the cusp and the base (Pl. 1, figs 1–4, cf. Rexroad 1967, pl. 4, figs 9–12); in this respect, it resembles the equivalent element in *W. bicostatus* (Branson and Mehl), as illustrated by Armstrong (1990, pl. 21, fig. 5). The bases of all the S elements are a little longer in the Chinese specimens than in typical *W. curvatus*, but the distribution of costae appears to be identical. In sample Shiqian 9 (Leijiatun Formation), some of the specimens of the P element (Pl. 2, figs 11–12) are relatively strongly curved so that the distal portion of the cusp becomes perpendicular to the upper edge of the cusp; in specimens of the same element from the Brassfield Formation, figured as *Acodus unicostatus* by Rexroad (1967, pl. 4, figs 13–16), this angle is 40–50°. In the same Chinese sample, specimens of the M element are erect, with long cusps (Pl. 2, figs 9–10, 13–14) and may show a deflection in the curvature of the anterior edge close to the antero-aboral corner, resulting in a short straight or slightly convex profile. The latter feature is also shown by some possible juvenile specimens from sample Shiqian 7 (Pl. 1, figs 24–25, 30–31); the elements from these two samples are assigned to *Walliserodus* aff. *curvatus*.

A small number of poorly preserved *Walliserodus* elements (Pl. 2, figs 1–8) were found in sample Shiqian-1 (top Kuanyinchiao Bed, regarded as uppermost Ordovician). These are slender, gently curved elements, bearing a resemblance to some of the elements figured as *Walliserodus debolti* (Rexroad) by Serpagli (1967, pl. 31, figs 5a–9b only). *Paltodus debolti* is a junior subjective synonym of *Acodus curvatus* (see Cooper 1975, p. 995), so Serpagli's specimens might now be referred to *W. curvatus*. The specimens from Shiqian-1, however, clearly differ from *W. curvatus* in their gentle curvature and are here recorded as *Walliserodus* sp. A.

Family DAPSILODONTIDAE Sweet, 1988

Genus CHENODONTOS gen. nov.

Derivation of name. After Chen Xu, one of the Chinese co-leaders of the Transhemisphere Telychian project; odontos, Gr., tooth.

Type species. Chenodontos makros sp. nov.

Diagnosis. Apparatus of laterally compressed, thin-walled coniform elements. P and S elements with long bases and strongly recurved cusps; P elements with acostate or weakly costate lateral faces, S elements multicostate with a broad, flat, smooth area bordering the lower edge of the base on each lateral face. M element geniculate.

Remarks. The elements of this genus broadly resemble those of *Dapsilodus*, but the geniculate M element and the multicostate S elements are distinctive. *Besselodus* also has a geniculate M element (Aldridge 1982), but the S elements are more gently curved and bear only a single costa on each lateral face; the acostate P elements have not been recognized in *Besselodus* or *Dapsilodus*. *Chenodontos* also lacks the characteristic oblique anterior striae shown by *Besselodus* (Aldridge 1982, pl. 44) and *Dapsilodus* (e.g. Cooper 1976, pl. 2, figs 11–12, 18–20). At present, only the type species is known.

A geniculate element is also included in the apparatus of the early Silurian species *Ansella mischa* Bischoff, 1997, from New South Wales, Australia. The specimens of this element illustrated by Bischoff (1997, fig. 3) are similar to those included here in *Chenodontos*, especially that in Pl. 3, figs 30–31, but the Chinese samples that contain the geniculate element lack any specimens resembling the rest of the elements of the *Ansella* apparatus. In general, the aboral edge of the geniculate element of *Chenodontos* is more gently convex than in illustrated specimens of *Ansella mischa*. A geniculate element was also described and figured, as *Walliserodus?* n. sp. B, by McCracken (1991a, p. 82, pl. 4, figs 1–4) from the Canadian Cordillera, northern Yukon Territory. The identity of any elements from the same apparatus is unknown, but, as noted by McCracken (1991a, p. 82), similar geniculate elements are unknown in species of *Walliserodus*. It is possible that the Canadian specimens represent a species of *Chenodontos*; the aboral edge is more similar to the Chinese specimens than to the *Ansella* elements from Australia.

Chenodontos makros sp. nov.
Plate 3, figures 1–31

?1983 *Distacodus obliquicostatus* Branson et Mehl, 1933;
Zhou and Zhai, p. 273, pl. 65, fig. 21a–b.

Derivation of name. Gr., long, in reference to the long bases of the P and S elements.

Holotype. Specimen NIGPAS 149663 (Pl. 3, figs 18–19); typically costate Sb element.

Type locality and horizon. Sample Shiqian 8, Leijiatun Formation, Leijiatun section, Shiqian County, Guizhou Province.

Diagnosis. As for genus.

Material. Pa, 1; Pb, 2; M, 6; Sa?, 2; Sb, 13, Sc, 14.

Description. Pa element long-based, with a recurved cusp. Element slightly twisted, with a faint costa distally along the midline of each lateral face; faces otherwise smooth. Laterally compressed, with sharp anterior and posterior edges; basal cavity tip situated close to point of maximum curvature.

Pb element with very long base and short, recurved cusp. Lateral faces acostate, anterior and posterior edges sharp. Entire element strongly laterally compressed, with anterior and posterior margins very flat. Basal cavity extends to tip near point of maximum curvature.

M element geniculate, with a short base and longer, slightly twisted cusp, broken on all specimens. Cusp with convex outer face; inner face with axial bulge that extends into a flaring inner face on the base. Upper edge of base convex, forming a tight v-shaped angle with cusp, aboral edge of base convex or gently sigmoidal. Basal cavity wide and shallow, with tip close to point of maximum curvature on the anterior edge. A narrow wrinkle zone apparent parallel to aboral edge, and there is a tiny projection at antero-basal corner (Pl. 3, fig. 28).

S elements multicostate and intergrading morphologically. No truly symmetrical Sa element can be distinguished, but a laterally compressed form with a similar pattern of costae on each lateral face (Pl. 3, figs 5–6, 11–12) probably represents this location. This element has two prominent costae on each face with minor costae developed anterior to these; the main costae on one face join distally, on the other they remain subparallel. Sb elements variable, but all long-based with at least two prominent costae on each lateral face, sometimes with three or four; these costae commonly merge towards cusp. Some specimens with a variably developed antero-lateral costa on inner side (Pl. 3, figs 10, 15, 25) and posterior edge that may bear two parallel costae. On other specimens anterior margin acostate, but bent strongly inwards (Pl. 3, fig. 18). Sc element with shorter base and strong laterally directed costa on inner anterior edge; inner face with one or two costae towards the posterior edge, outer face with a weak costa close to posterior edge on the base, becoming asymptotic to posterior edge of cusp. On all S elements, costae do not extend to aboral edge. Basal cavity widely flared inwards on Sb and Sc elements; aboral edges usually damaged but appear to have been straight. Basal cavity tip situated axially, proximal to point of maximum recurvature.

Occurrence. Leijiatun Formation, Leijiatun section, Shiqian County, Guizhou; Yangpowan Member, Ningqiang Formation, Yushitan section, Ningqiang, Shaanxi (sample Ningqiang 3).

Order PROTOPANDERODONTIDA Sweet, 1988
Family PSEUDOONEOTODIDAE fam. nov.

Remarks. *Pseudooneotodus* was included with equivocation by Sweet (1988) in the family Protopanderodontidae. As noted by Aldridge and Smith (1993), this is a large family group, probably encompassing several lineages, perhaps with a common ancestor within *Semiacontiodus*. The apparatus of *Pseudooneotodus* is uncertain, but it is very unlikely that it fits within this grouping. Some of the taxa included by Sweet in the Protopanderodontidae had previously been accommodated in a separate family, the Oneotodontidae Miller (*in* Robison 1981), and Aldridge and Smith (1993) recognized this separation. The Oneotodontidae, as originally established, incorporated multimembrate apparatuses

---

EXPLANATION OF PLATE 1

Figs 1–23. *Walliserodus curvatus* (Branson and Branson, 1947). Xiangshuyuan Formation, Leijiatun Section, Shiqian County, Guizhou, Sample Shiqian 5. 1–2, 149623, M element, inner and outer lateral views. 3–4, 149624, M element, outer and inner lateral views. 5–6, 149625, P element, inner and outer lateral views. 7–8, 149626, Sb element, inner and outer lateral views. 9–10, 149627, P element, inner and outer lateral views. 11–12, 149628, Sb element, lateral views. 13–14, 149629, Sa element, lateral views. 15–16, 149630, Sb element, lateral views. 17–18, 149631, Sb element, lateral views. 19–20, 149632, Sb element, lateral views. 21, 149633, Sa element, lateral view showing regenerated cusp. 22–23, 149634, Sc element, outer and inner lateral views.

Figs 24–33. *Walliserodus* aff. *curvatus* (Branson and Branson, 1947). Leijiatun Formation, Leijiatun Section, Shiqian County, Guizhou, Sample Shiqian 7. 24–25, 149635, M element, outer and inner lateral views. 26–27, 149636, Sb element, lateral views. 28–29, 149637, Sc element, outer and inner lateral views. 30–31, 149638, M element, inner and outer lateral views. 32–33, 149639, P element, outer and inner lateral views.

All figures ×80.

PLATE 1

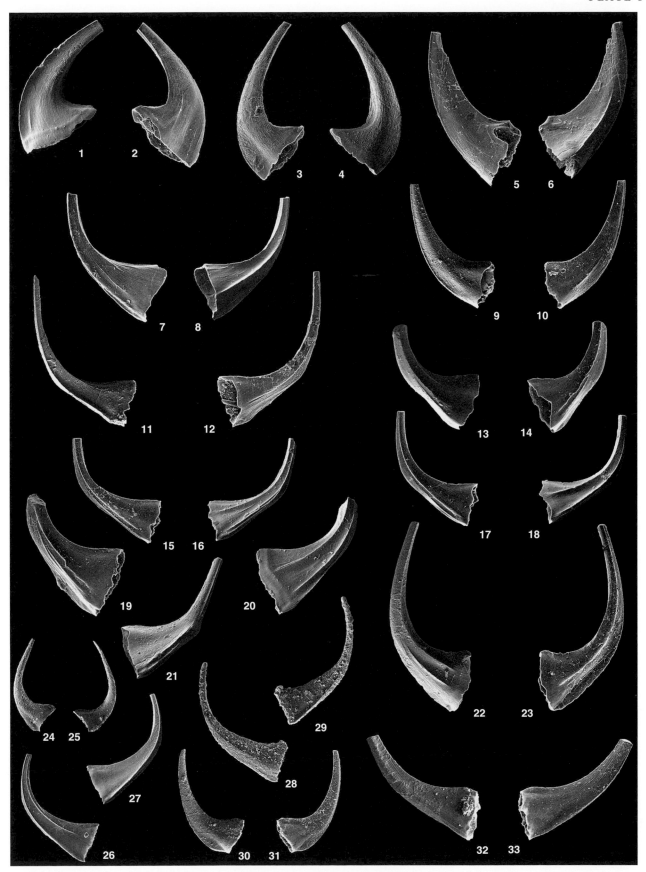

WANG and ALDRIDGE, *Walliserodus*

in which one element lacks costae and the others bear multiple lateral or posterior costae. *Pseudooneotodus* does not fit this diagnosis and was not included in the family by Miller (*in* Robison 1981). Dzik (1991) questionably assigned the genus to the family Fryxellodontidae Miller (*in* Robison 1981), but *Fryxellodontus* is interpreted to have a multimembrate apparatus that includes a serrate element, and there is no evidence that any similarity between the two genera is other than superficial. Sansom (1996) also noted gross morphological similarities between early species of *Pseudooneotodus* and elements of the genus *Polonodus* Dzik, but stated that further work was needed to confirm any relationship. Aldridge and Smith (1993) allocated *Pseudooneotodus* to an unnamed family (fam. nov. 5), which also possibly includes *Fungulodus* Gagiev, 1979 (= *Mitrellataxis* Chauff and Price, 1980). We follow this separation and give the name Pseudooneotodidae to the family that includes *Pseudooneotodus*.

Dzik (1991) included the Fryxellodontidae, and, therefore, *Pseudooneotodus*, in the Order Panderodontida, which he regarded as distinguished by apparatuses lacking a symmetrical medial element. Some specimens of the single-tipped squat conical element of *Pseudooneotodus*, however, are bilaterally symmetrical and might have occupied a medial position. In any event, we follow here Sweet's (1988) concept of the Panderodontida, which is restricted to taxa with a panderodontid furrow. *Pseudooneotodus* elements are not furrowed, so we tentatively include the family Pseudooneotodidae in the Order Protopanderodontida.

More recently, Dzik (2006) named a new family Jablonnodontidae, within the Order Prioniodontida, to accommodate the Famennian coniform genera *Mitrella-* *taxis* and *Jablonnodus* Dzik, 2006. He suggested that these Late Devonian coniforms might represent either an icriodontid lineage with completely reduced P elements or survivors of Ordovician protopanderodontids, separated by a Lazarus gap. A third possibility is that there might be a relationship between the Family Jablonnodontidae and the Family Pseudooneodontidae, although there is currently no direct evidence to support such a link.

Genus PSEUDOONEOTODUS Drygant, 1974

Type species. Oneotodus? beckmanni Bischoff and Sanneman, 1958.

Remarks. There have been various interpretations of the apparatus of *Pseudooneotodus*, and the number and types of element possessed by an individual remain in doubt. Barrick (1977) reconstructed apparatuses of *P. bicornis* Drygant, 1974, and *P. tricornis* Drygant, 1974, each consisting of three different elements: a two-tipped or a three-tipped element respectively, plus a one-tipped squat element and a slender conical element. He also speculated that the apparatus of the type species, *P. beckmanni*, might have comprised a single-tipped squat element and slender conical elements. Bischoff (1986) did not accept these reconstructions, because he did not find the single-tipped squat and slender conical elements in co-occurrence with the two-tipped and three-tipped elements in his collections from New South Wales. He considered the apparatus of *Pseudooneotodus* to be unimembrate. Armstrong (1990), however, reinforced the evidence for a trimembrate apparatus for *P. bicornis* and

---

EXPLANATION OF PLATE 2

Figs 1–8. *Walliserodus* sp. A. Kuanyinchiao Bed, Leijiatun Section, Shiqian County, Guizhou, Sample Shiqian-1. 1-2, 149640, M element, inner and outer lateral views. 3–4, 149641, P element, inner and outer lateral views. 5–6, 149642, Sc element, lateral views. 7–8, 149643, Sa element, lateral views.

Figs 9–18. *Walliserodus* aff. *curvatus* (Branson and Branson, 1947). Leijiatun Formation, Leijiatun Section, Shiqian County, Guizhou, Sample Shiqian 9. 9–10, 149644, M element, outer and inner lateral views. 11–12, 149645, P element, inner and outer lateral views. 13–14, 149646, M element, inner and outer lateral views. 15–16, 149647, Sc element, lateral views. 17–18, 149648, Sb element, lateral views.

Figs 19–24. *Pseudooneotodus beckmanni* (Bischoff and Sannemann, 1958). Shenxuanyi Member, Xuanhe Section, Guangyuan County, Sichuan, Sample Xuanhe 1. 19–20, 149649, lateral and oral views. 21–22, 149650, lateral and oral views. 23–24, 149651, ?coniform element, oral and lateral views.

Figs 25–26. *Pseudooneotodus beckmanni* (Bischoff and Sannemann, 1958). Shenxuanyi Member, Xuanhe Section, Guangyuan County, Sichuan, Sample Xuanhe 2. 149652, lateral and oral views.

Figs 27–28. *Pseudooneotodus?* sp. Daluzhai Formation, Huanggexi Section, Daguan County, Yunnan, Sample TT 1169. 149653, oral view and lateral view of cusp tip.

Fig. 29. *Decoriconus fragilis* (Branson and Mehl, 1933a). Xiangshuyuan Formation, Leijiatun Section, Shiqian County, Guizhou, Sample Shiqian 3. 149654, Sb? element, inner lateral view.

Figs 1–27, ×80, figs 28-29, ×120.

PLATE 2

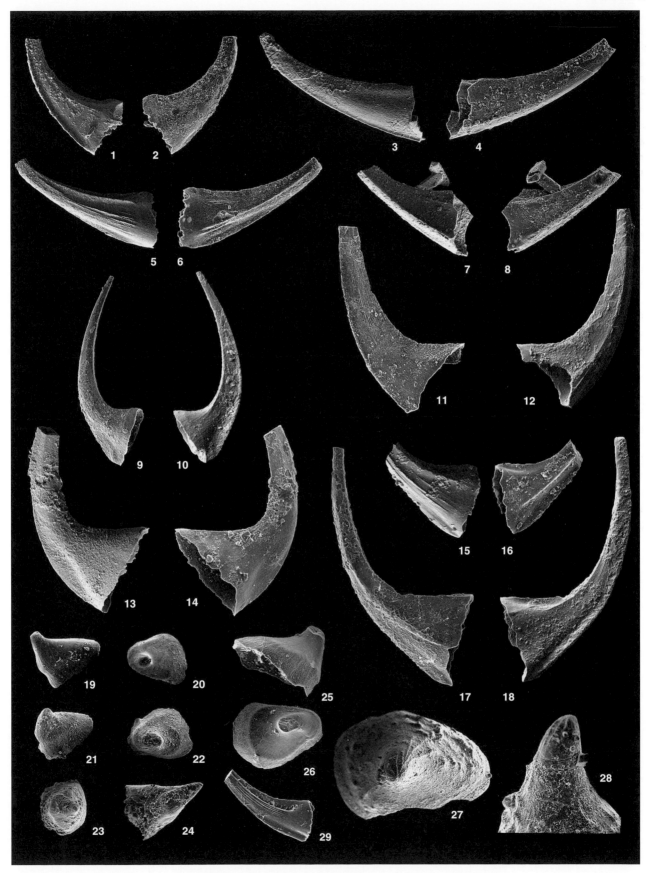

WANG and ALDRIDGE, *Walliserodus, Pseudooneotodus*

P. tricornis on the basis of collections from north Greenland, regarding unidenticulate elements to be vicarious. More recently, in a study of large numbers of specimens from Sardinia and the Carnic Alps, Corradini (2001, 2008) has found that two-tipped specimens of *P. bicornis* do not co-occur regularly with one-tipped elements, and that the slender conical element was never recovered together with the squat elements; he, therefore, concluded that the apparatuses of *P. beckmanni* and *P. bicornis* contained elements of only a single morphological type.

Interpretation is even more difficult when one considers the many collections that contain only single-tipped squat elements. These have commonly all been referred to *P. beckmanni*, and Cooper (1977, p. 1069) used this name to accommodate all such material from Middle Ordovician to Early Devonian in age. Bischoff (1986), however, restudied material from the type area and reported that the outline of the basal margin is always subtriangular; he, therefore, proposed that specimens with other basal outlines should be assigned to other species. He also noted that the suite of specimens with subtriangular bases encompassed forms with right and left asymmetry and with bilateral symmetry, suggestive of variation within the apparatus. Barrick (1977) had speculated that the apparatus of *P. beckmanni* may have comprised a single-tipped squat element and slender conical elements, but Bischoff (1986) reported no evidence for this. Armstrong (1990) considered that published evidence and his own Greenland collections supported the concept of a unimembrate apparatus for *P. beckmanni*, and Zhang and Barnes (2002) further supported this on the evidence from their collections from Anticosti Island, Québec. Corradini (2008) also accepted a unimembrate structure, but included specimens with a wide range of basal outlines, from subtriangular to ovoid and rarely subrectangular, within his concept of the species.

Whatever the number of element types in the various *Pseudooneotodus* apparatuses, the actual number of elements is totally unknown. It will probably need the discovery of natural assemblages to resolve the questions surrounding the apparatus composition and architecture of this genus.

Pseudooneotodus beckmanni (Bischoff and Sannemann, 1958)
Plate 2, figures 19–26

1958 *Oneotodus? beckmanni* Bischoff and Sannemann, p. 98, pl. 15, figs 22–25.
2008 *Pseudooneotodus beckmanni* (Bischoff and Sannemann, 1958); Corradini, p. 142, pl. 1, figs 1–7 (with synonymy to 2006).

Material. 16 specimens, plus additional material from TT samples.

Remarks. Because of the uncertainty surrounding the circumscription of this species, we have not attempted to provide a critical synonymy list. As noted above, Bischoff (1986) studied material from the type area and considered that the name should be restricted to specimens with a subtriangular outline to the base. Most of our specimens accord with this (Pl. 2, figs 19–22, 25–26), but we have also found in association a more conical element with a subcircular base (Pl. 2, figs 23–24) that may belong to the same apparatus. This specimen resembles those assigned by Bischoff (1986, pp. 235–237, pl. 27, figs 13–17) to his new species *P. boreensis*, but lacks the thickened basal rim. However, we have too few specimens in our collections to provide a meaningful contribution to the debate on the apparatus composition and element morphology of *P. beckmanni*.

Occurrence. In small numbers: Xiangshuyuan Formation and upper member, Xiushan Formation, Leijiatun section, Shiqian County, Guizhou; Yangpowan Member, Ningqiang Formation,

---

EXPLANATION OF PLATE 3

Figs 1–15, 18–19, 26–29. *Chenodontos makros* gen. et sp. nov. Leijiatun Formation, Leijiatun Section, Shiqian County, Guizhou, Sample Shiqian 8. 1–2, 149655, Pb element, lateral views. 3–4, 149656, Pa element, lateral views. 5–6, 149657, Sa? element, lateral views. 7–8, 149658, Sc element, outer and inner lateral views. 9–10, 149659, Sb element, outer and inner lateral views. 11–12, 149660, Sa? element, lateral views. 13, 149661, Sb element, inner lateral view. 14–15, 149662, Sb element, outer and inner lateral views. 18–19, 149663, Sb element, inner and outer lateral views (holotype). 26, 149664, M element, oblique inner lateral view. 27–28, 149665, M element, inner lateral view and close-up of basal margin to show wrinkle zone and antero-basal projection. 29, 149666, M element, inner lateral view.

Figs 16–17, 20–25, 30–31. *Chenodontos makros* gen. et sp. nov. Leijiatun Formation, Leijiatun Section, Shiqian County, Guizhou, Sample Shiqian 9. 16–17, 149667, Sc element, inner and outer lateral views. 20–21, 149668, Sc element, outer and inner lateral views. 22–23, 149669, Sb element, outer and inner lateral views. 24–25, 1496670, Sb element, outer and inner lateral views. 30–31, 149671, M element, outer and inner lateral views.

All figures ×80, except fig. 28, ×280.

PLATE 3

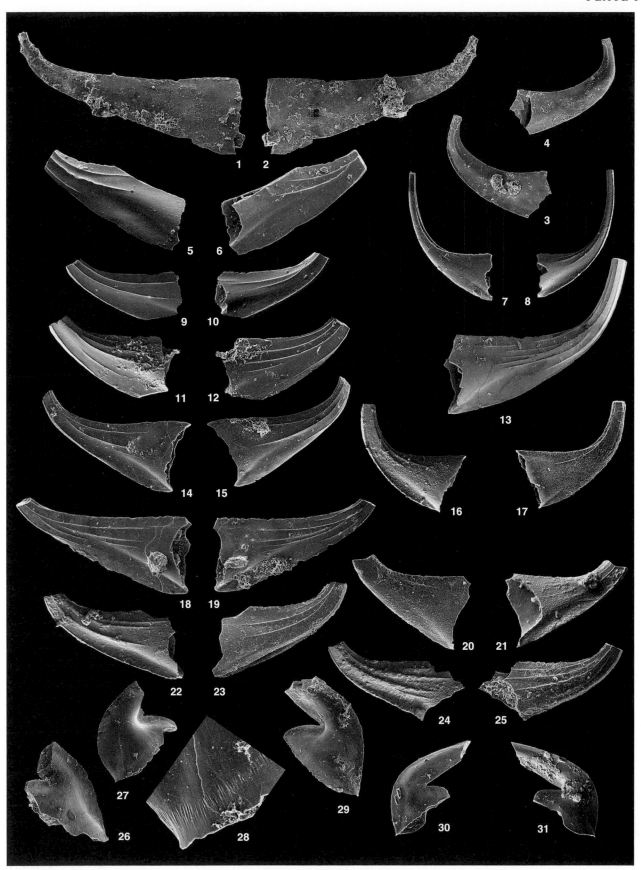

WANG and ALDRIDGE, *Chenodontos*

Yushitan section, Ningqiang, Shaanxi; Shenxuanyi Member, Xuanhe section, Guangyuan, Sichuan.

Pseudooneotodus? sp.
Plate 2, figures 27–28

Description. A single specimen of a single-tipped slender conical, posteriorly curved element with an ovoidal basal outline. Tip circular in cross-section, with radial costae extending downwards from the apex, but not reaching basal margin of unit. Small ridges connect the radial costae to form an irregular network in the apical area; these become nodose towards aboral margin and form incomplete ovoidal rings parallel to basal outline.

Remarks. The overall shape of this element is very similar to some specimens of *Pseudooneotodus*, for example *P.* sp. b of Bischoff (1986, pl. 27, fig. 39). Cooper (1977, p. 1069. pl. 2, figs 12–13) illustrated a similar element as *Pseudooneotodus* n. sp.; this specimen has a comparable basal outline to ours, but is more strongly nodose. Mabillard and Aldridge (1983, p. 32, pl. 1, figs 10–11) tentatively included a similar ornamented conical element in the apparatus of *Apsidognathus ruginosus*, but the presence of this type of element in *Apsidognathus* apparatuses has not been confirmed by other studies.

Occurrence. Daluzhai Formation, Hunggexi section, Daguan County, Yunnan.

Order PANDERODONTIDA Sweet, 1988
Family PANDERODONTIDAE Lindström, 1970

Genus PANDERODUS Ethington, 1959

Type species. Paltodus unicostatus Branson and Mehl, 1933*a*, p. 42.

Remarks. Elements referable to *Panderodus* are very variable in the collections we have studied, and at least five species have been distinguished. Three of these, *P. panderi* (Stauffer, 1940), *P. serratus* Rexroad, 1967, and *P. unicostatus* (Branson and Mehl, 1933*a*) are reasonably well known; a fourth is a new species, *P. amplicostatus*. The fifth also appears to be new and is recorded here as *Panderodus* sp. nov. A. *P. panderi* has been described in detail (as *P. recurvatus*) by Zhang and Barnes (2002, p. 31), based on specimens from the Early Silurian of Québec. These authors reconstructed an apparatus that differs in some respects from that identifiable in the Chinese collections. For example, the aequaliform element from China sometimes lacks costae and is the only element with furrows on both sides; Zhang and Barnes (2002) additionally recognized other double-furrowed elements, which they termed symmetrical acostatiform and asymmetrical acostatiform. Other authors (e.g. Barrick 1977) have not identified these elements, and it may be that the Canadian apparatus is atypical. Zhang and Barnes (2002, p. 32) suggested that the asymmetrical acostatiform element may occupy the position taken by tortiform elements in other *Panderodus* apparatuses, but we find a clear tortiform in our material (Pl. 5, figs 8–9); Sansom *et al.* (1994, text-fig. 7) also found a tortiform element in *P. panderi* from our collections from Leijiatun (sample Shiqian 18). Zhang and Barnes (2002) further noted that their apparatus did not match the architecture deduced by Sansom *et al.* (1994) for *Panderodus* and that, if their reconstruction is correct, this species should perhaps be reassigned to a new genus. The apparatus of *P. panderi*, as we recognize it, fully accords with the Sansom *et al.* (1994) model, which does not support an assignment to a different genus. It will be interesting to see whether future authors discover the acostatiform elements in their collections.

Although *Panderodus unicostatus* has been widely identified in Early Silurian collections (see synonymy list in Zhang and Barnes 2002, p. 32), it is one of a group of *Panderodus* species that are readily confused (see, for example, the discussion by Jeppsson 1983), the others including *P. acostatus* (Branson and Branson, 1947), *P. equicostatus* (Rhodes, 1953) and *P. gracilis* (Branson and Mehl, 1933*c*). It is possible that *P. acostatus* and *P. equicostatus* are synonyms, and the same applies to *P. gracilis* and *P. unicostatus*, although Sansom *et al.* (1994, text-fig. 7) maintained them as distinct species; Zhang and Barnes (2002, p. 33) suggested that *P. unicostatus* and *P. acostatus* might be synonymous. Key differences between these four species, as shown in the diagrams given by Sansom *et al.* (1994, text-fig. 7), occur in the truncatiform and

---

EXPLANATION OF PLATE 4

Figs 1–21. *Panderodus amplicostatus* sp. nov. Shenxuanyi Member, Ningqiang Formation, Yushitan Section, Ningqiang County, Shaanxi, Sample Ningqiang 2. 1–2, 149672, qt element, lateral views. 3–4, 149673, pf element, lateral views. 5–6, 149674, pt element, lateral views. 7–9, 149675, qa element, lateral views and close-up of basal wrinkle zone (holotype). 10–11, 149676, qa element, lateral views. 12–13, 149677, pt element, lateral views. 14–15, 149678, qg element, lateral views. 16–17, 149679, qg element, lateral views. 18–19, 149680, qg element, lateral views. 20–21, 149681, qg element, lateral views.

All figures ×80, except fig. 9, ×160.

PLATE 4

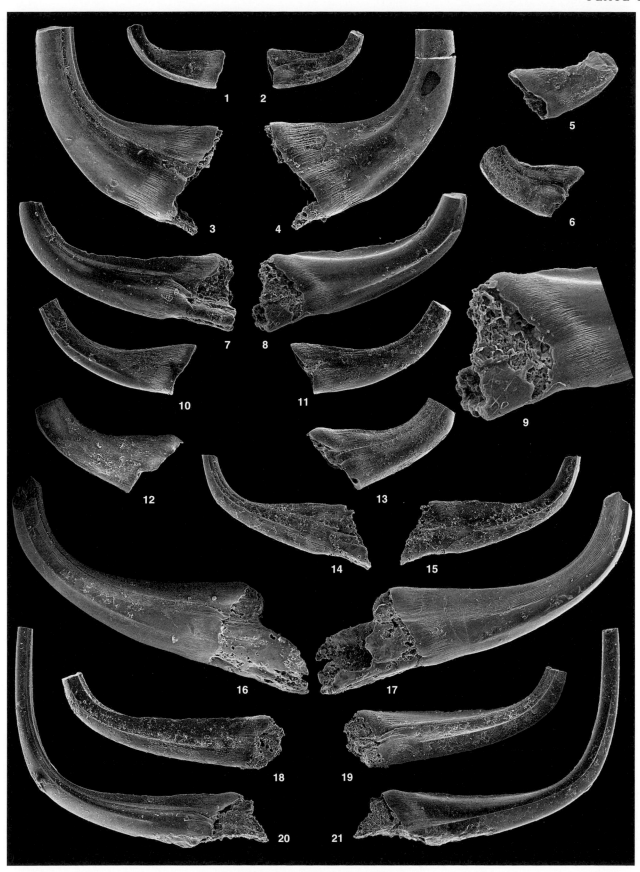

WANG and ALDRIDGE, *Panderodus*

aequaliform elements, which are more slender in *P. gracilis* and shorter-based in *P. acostatus*. The q (= S) elements are also more recurved in *P. acostatus*, and the falciform element is less strongly curved. In the apparatus we illustrate here from Shiqian 20 (Pl. 6, figs 1–18), the characteristics are all consistent with *P. unicostatus*, although the falciform (pf) element is more strongly recurved than in most previously illustrated specimens; it is possible that more than one species within this group is represented in our collections. *Panderodus serratus* (Pl. 5, figs 18–28) is similar to *P. unicostatus*, but the arcuatiform (qa) element has a clearly serrated edge on the concave margin.

The taxon distinguished here as *Panderodus* sp. nov. A (Pl. 7, figs 1–18) is similar to *P. panderi*, but is characterized by a very distinctive tortiform element (Pl. 7, figs 9–10), with a long base and rounded subquadrate cross-section to the cusp. Elements are most abundant in sample Xuanhe 1, but they are somewhat recrystallized and overgrown. For this reason, and because there are numerous elements of *Panderodus* in the sample rendering it difficult to be sure which belong to the same apparatus, the new species is retained in open nomenclature. One feature apparent on the elements of this apparatus, where they are not too recrystallized, is the presence of well-developed longitudinal striae on all faces.

The description below uses the morphological terminology and locational notation of Sansom *et al.* (1994).

Panderodus amplicostatus sp. nov.
Plate 4, figures 1–21

Derivation of name. L., amplus, large and costatus, ribbed.

Holotype. Specimen NIGPAS 149675 (Pl. 4, figs 7–9); arcuatiform (qa) element.

Type locality and horizon. Sample Ningqiang 2, Ningqiang Formation, Yushitan section, Ningqiang, Shaanxi Province.

Diagnosis. All elements robust, p and q elements with well-developed costae. pf element with relatively short, high base and strongly recurved cusp; q elements with long bases and recurved cusps. White matter continues to the tips of all elements.

Material. pt, 14; pf, 52; qt, 4; qg, 60; qa, 41.

Description. Tortiform (pt) element generally small, relatively short-based, twisted, with a smooth, unfurrowed inner face bearing an antero-lateral costa, sometimes prominent, at the anterior margin; costa paralleled on its posterior side by a shallow groove. Outer face with a broader, rounder costa between furrow and posterior margin.

Falciform (pf) element with a short, high base and erect cusp. Lower part of anterior edge keeled, costae also developed on furrowed face close to posterior edge and commonly between the furrow and anterior margin.

Aeqaliform (ae) element not discriminated.

Truncatiform (qt) element small, with a long base, cusp slightly twisted. Each lateral face bears a prominent costa, closer to anterior margin on furrowed face. Anterior edge rounded, posterior edge sharper.

Graciliform (qg) elements robust with long bases and costate lateral faces. Costae pronounced, at least one on each face; the main costa on each face may be symmetrically (Pl. 4, figs 14–15, 18–21) or asymmetrically (Pl. 4, figs 16–17) disposed; bases may be relatively high (Pl. 4, figs 14–17) or low (Pl. 4, figs 18–21).

Arcuatiform (qa) element with long base and slightly twisted cusp. Prominent costa developed subparallel to posterior edge on the inner, unfurrowed face. Upper margin of base in larger specimens developed into a knife-like sharp edge.

All elements show a well-developed wrinkle zone parallel to the aboral margin and a basal funnel is commonly preserved. Cusp broken away on almost all specimens, but where intact shows that white matter occurs to the apex.

Remarks. Panderodus amplicostatus is a distinctive species. In the strongly costate character of the elements, it is comparable with *P. fuelneri* (Glenister, 1957), but the p and q elements of the latter are much more gently curved, and the pf element has a longer base (see Sweet, 1979, fig. 7.1, 8, 11–14, 17–18, 22; Sansom *et al.* 1994,

EXPLANATION OF PLATE 5

Figs 1–17. *Panderodus panderi* (Stauffer, 1940). Upper member, Xiushan Formation, Leijiatun Section, Shiqian County, Guizhou, Sample Shiqian 17. 1–3, 149682, pf element, lateral views and close-up of basal wrinkle zone. 4–5, 149683, qt element, lateral views. 6–7, 149684, ae element, lateral views. 8–9, 149685, pt element, lateral views. 10–11, 149686, qg element, lateral views. 12–13, 149687, qg element, lateral views. 14–15, 149688, qg element, lateral views. 16–17, 149689, qa element, lateral views.

Figs 18–32. *Panderodus serratus* Rexroad, 1967. Leijiatun Formation, Leijiatun Section, Shiqian County, Guizhou, Sample Shiqian 7. 18–19, 149690, pf element, lateral views. 20–21, 149691. qt element, lateral views. 22–23, 149692, pt element, lateral views. 24–25, 149693, qa element, lateral views. 26–28, 149694, qa element lateral views and close-up to show serrated edge. 29–30, 149695, qg element, lateral views. 31–32, 149696, qg? element, lateral views.

All figures ×80, except fig. 3, ×360, fig. 28, ×750.

PLATE 5

WANG and ALDRIDGE, *Panderodus*

text-fig. 7). There is also some similarity to *P. greenland-ensis* Armstrong, 1990, which also has robust long-based elements with strong costae (Armstrong, 1990, p. 102, fig. 33), but the suite of type elements of that species are all very slender, and the wrinkle zone is much longer; the white matter characteristically does not extend to the tip of the cusp. Specimens of *P.* aff. *greenlandensis* from Greenland (Aldridge 1979, pl. 2, figs 23–30; refigured Armstrong 1990, pl. 15, figs 1–8) have much shorter bases and a long wrinkle zone that extends to the aboral edge. Specimens from Greenland referred to *P.* cf. *fuelneri* by Armstrong (1990, pl. 15, figs 9–14) all have very short bases.

Occurrence. Yangpowan and Shenxuanyi members, Ningqiang Formation, Yushitan section, Ningqiang, Shaanxi; Shenxuanyi Member, Xuanhe section, Guangyuan, Sichuan.

Genus PSEUDOBELODELLA Armstrong, 1990

Type species. Pseudobelodella silurica Armstrong, 1990, p. 111, by original designation.

Remarks. Pseudobelodella is tentatively placed in the Panderodontidae as all elements show a clear furrow on one lateral face. However, we have not been able to identify the suite of elements recognized in *Panderodus*; there is no double-furrowed, aequaliform element, and we cannot distinguish tortiform or truncatiform elements. We have, therefore, not attempted to apply the locational notation developed for *Panderodus* by Sansom *et al.* (1994) to *Pseudobelodella*.

Pseudobelodella spatha (Zhou, Zhai and Xian, 1981)
Plate 8, figures 1–32

*1981 *Belodella spatha* Zhou et al., p. 130, pl. 2, figs 20–21.
1992 *Belodella* sp. A. Qian *in* Jin et al., pl. 2, fig. 12a–b.
p 1993 *Belodella* sp.nov. Xia, p. 207, pl. 1, figs 2–3, 5 [only, *non* figs 4, 6, 14].

Diagnosis. P? element compressed with a relatively high base; upper edge of base bears several erect denticles. S elements with very long bases and inwardly bowed cusps; upper edge of base with numerous, closely spaced denticles.

Material. P?, 20; S, 128; plus additional material from TT samples.

Description. P? element with relatively high base and broad, short, slightly proclined cusp. Lower margin of element continuously curved from base to apex. Upper edge of base with 8–16 small, erect, closely packed denticles, very short and completely fused near the aboral margin, becoming tall and slender in central part of the denticle row and larger and broader near beginning of cusp. Lower edge of base pinched near aboral margin (Pl. 8, fig. 8). Entire aboral surface excavated, with cavity flaring towards the lower corner.

S elements either long-based and gently bowed (equivalent to qg elements in *Panderodus*) or shorter-based and more strongly bowed (equivalent to qa elements in *Panderodus*). Bowing commonly, but not always, away from furrowed face. Long-based elements have a short proclined cusp, slightly inwardly bowed, and a costa on each lateral face, more or less symmetrically disposed. Each costa runs from lower basal corner, curving gently away from lower edge of base and then back to meet anterior edge at beginning of cusp; area between costa and lower edge of base flat and lenticular in shape. Upper margin of base bears numerous short, fused, apically inclined, sharp denticles, which increase in size a little towards apex. Close to aboral edge the upper margin is adenticulate. Lower edge of base very gently convex, the curvature increasing as base runs into cusp. Basal cavity deep with triangular apex, terminating at approximately half the length of the element. Short-based S elements are much more strongly bowed and curved and strongly compressed; some specimens have shorter bases than others, but all distinctly shorter and higher than the long-based forms. Base high, its upper margin near aboral edge adenticulate, forming sharp corner with aboral edge. Upper edge of base bears 13–18 fused denticles, which increase in size towards apex; the most distal is the largest. Basal cavity long, terminating at about half the length of the element.

All elements small, with straight aboral edges and a distinct wrinkle zone parallel to the aboral margin (Pl. 8, fig. 8).

Remarks. Fordham (1991, p. 26) included 'Belodella spatha' in synonymy with *Panderodus gracilis*, but the specimen illustrated by Zhou *et al.* (1981) appears to belong to the distinctive apparatus assigned here to *Pseudobelodella spatha*.

EXPLANATION OF PLATE 6

Figs 1–18. *Panderodus unicostatus* (Branson and Mehl, 1933a). Upper member, Xiushan Formation, Leijiatun Section, Shiqian County, Guizhou, Sample Shiqian 20. 1–2, 149697, qt element, lateral views. 3–4, 149698, ae element, lateral views. 5–6, 149699, pf element, lateral views. 7–8, 149700, qa element, lateral views. 9–10, 149701, pt element, lateral views. 11–12, 149702, pt element, lateral views. 13–14, 149703, qg? element, lateral views. 15–16, 149704, qg element, lateral views. 17–18, 149705, qg element, lateral views.

All figures ×80.

PLATE 6

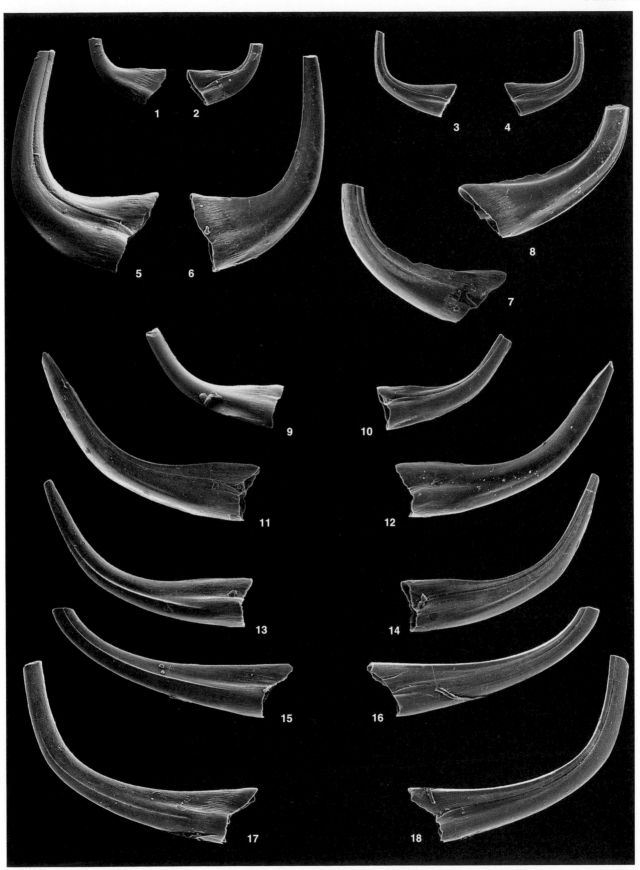

WANG and ALDRIDGE, *Panderodus*

Pseudobelodella spatha differs from *Pseudobelodella silurica* Armstrong, 1990, in several respects. The P? element of the latter is shorter-based and bears only four to six denticles (Armstrong 1990, p. 111, as 'sq element'). The lateral costae on S elements of *P. silurica* are parallel to the lower edge of the base, and the characteristic flat, lenticular area of *P. spatha* is not developed.

There are some small differences in different populations of *P. spatha* from China. Denticulation on specimens from Ningqiang (Pl. 8, figs 1–17) appears generally to be a little less erect than on those from Leijiatun (Pl. 8, figs 18–31), and the short-based S elements are a little longer and lower. These differences are not considered significant enough to warrant taxonomic differentiation. A pair of fused S elements from Xuanhe 8 (Pl. 8, fig. 32) shows a difference in denticulation on the upper margin of each specimen, with one displaying even denticulation and the other with alternating denticles clearly developed in the central part. Alternating denticulation has not been recognized in other Chinese collections of *Pseudobelodella*, and this pair might represent a distinct species. However, other specimens in the same sample have relatively even denticulation, so no taxonomic discrimination is made at present.

Occurrence. Xiangshuyuan Formation and lower part of upper member, Xiushan Formation, Leijiatun section, Shiqian, Guizhou; Yangpowan and Shenxuanyi members, Ningqiang Formation, Yushitan section, Ningqiang, Shaanxi; Shenxuanyi Member, Xuanhe section, Guangyuan, Sichuan.

Division PRIONIODONTIDA Dzik, 1976

Remarks. The order Prioniodontida, as traditionally conceived, is paraphyletic (see Donoghue *et al.* 2000, fig. 2; Donoghue *et al.* 2008). Following the preliminary cladistic analysis of phylogenetic relationships within the non-coniform conodonts by Donoghue *et al.* (2008), we include all of these taxa within the Division Prioniodontida.

There is only limited evidence for the apparatus composition and architecture of the more basal prioniodontid conodonts (the traditional Order Prioniodontida). The principal direct evidence comes from several hundred preserved apparatuses of the balognathid genus *Promissum* Kovács-Endrödy (*in* Theron and Kovács-Endrödy 1986), from the Upper Ordovician Soom Shale of South Africa. *Promissum* has an apparatus of 19 elements, comprising paired P_1, P_2, P_3, P_4, M and S_{1-4} elements and an axial S_0 element (Aldridge *et al.* 1995; Purnell *et al.* 2000). The extent to which this apparatus structure is representative of early prioniodontids is uncertain, particularly as completely preserved apparatuses of *Phragmodus* Branson and Mehl, 1933*b*, a genus also included by Sweet (1988) in the Prioniodontida, show a 15-element arrangement similar to that of the ozarkodinids (Repetski *et al.* 1998). Some prioniodontid apparatuses have been reconstructed to include three pairs of P elements (Pa, Pb, Pc), for example *Pterospathodus* Walliser (Männik and Aldridge 1989), *Pranognathus* Männik and Aldridge (Männik and Aldridge 1989), *Astropentagnathus* Mostler (Armstrong 1990) and *Coryssognathus* Link and Druce (Miller and Aldridge 1993). Direct support for this is found in unpublished collections of complete assemblages of a new Late Ordovician balognathid taxon from the same localities as the *Promissum* apparatuses in South Africa. This apparatus has pairs of Pa, Pb and Pc elements, a pair of M elements, and an array of nine S elements (homologous to the S_{0-4} elements of *Promissum*).

Family DISTOMODONTIDAE Klapper, 1981
Genus CORYSSOGNATHUS Link and Druce, 1972

?1980 *Dentacodina* Wang, p. 370.

Type species. Coryssognathus dentatus Link and Druce, 1972, subjective junior synonym of *Cordylodus? dubius* Rhodes, 1953.

Diagnosis. See Miller and Aldridge (1993, p. 242).

Remarks. The apparatus of *Coryssognathus* was reconstructed by Miller and Aldridge (1993) to contain Pa, Pb, Pc, M, Sa/Sb and Sc elements. The type species has been widely reported from Late Silurian collections (see summary in Miller and Aldridge 1993). Specimens very similar to the M and S elements of *C. dubius* (Rhodes) have been reported previously from the Early Silurian of Greenland

---

PLATE 7

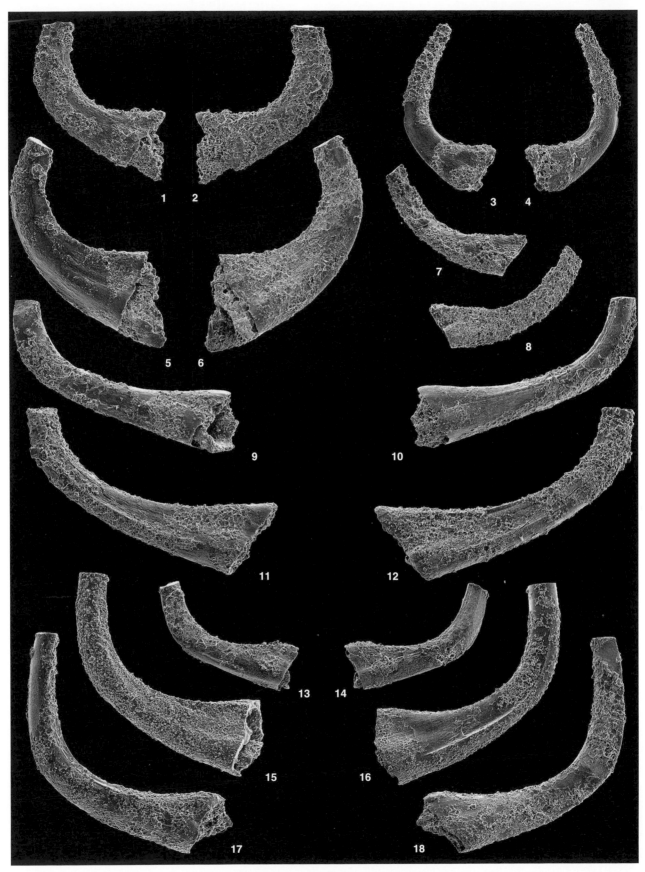

WANG and ALDRIDGE, *Panderodus*

(Armstrong 1990) and were referred to *Dentacodina*. Broadly similar specimens from the Early Silurian Coralliferous Group of Pembrokeshire, Wales, were referred to *Rotundacodina* Carls and Gandl by Mabillard and Aldridge (1983), and a small number of specimens from the underlying Skomer Volcanic Group of Pembrokehire were identified as *Coryssognathus*? sp. by Aldridge (2002). None of these Early Silurian collections contains Pa elements comparable to those of *C. dubius*, although Armstrong (1990, p. 72) suggested that a broad-based coniform element occupied this position in his apparatus. In the Chinese material, we have recognized a scaphate Pa element comparable with that of Late Silurian *Coryssognathus*, but, whereas our collection contains relatively abundant specimens of the M and S elements, only a few specimens of the Pa have been recovered. This under-representation of Pa elements has been noted previously for Late Silurian collections of *Coryssognathus* (van den Boogaard 1990; Miller and Aldridge 1993). It may well be, therefore, that the collections from Wales and Greenland also represent species of *Coryssognathus* and that the characteristic Pa element has simply not been recovered.

Miller and Aldridge (1993) regarded Late Silurian elements previously referred to *Dentacodina* to belong to the apparatus of *Coryssognathus dubius* (Rhodes, 1953) and considered *Dentacodina* to be a possible junior synonym of *Coryssognathus*. However, the apparatus of the type species of *Dentacodina* is still unknown; until the nature of this is resolved, the synonymy must remain equivocal.

Coryssognathus shaannanensis (Ding and Li, 1985)
Plate 9, figures 1–27

1983 *Dentacodina dubia* (Rhodes, 1953); Zhou and Zhai, p. 272, pl. 66, fig. 27a–b.

1983 *Dentacodina multidentata* Wang, 1980; Zhou and Zhai, p. 272, pl. 67, fig. 17a–b.
1983 *Dentacodina trilinearis* Wang, 1980; Zhou and Zhai, p. 272, pl. 66, fig. 33a–b.
*1985 *Dentacodina shaannanensis* Ding and Li, p. 16, pl. 1, figs 24–25.
1987 *Distomodus*? cf. *dubius* (Rhodes); Ni, p. 402, pl. 62, fig. 17.
1987 *Dentacodina dubia* (Rhodes) 1953; An, p. 197, pl. 33, figs 17–19.
1987 *Dentacodina multidentata* Wang, 1980; An, p. 198, pl. 35, figs 10–11.
1990 *Dentacodina multidentata* Wang; An and Zheng, pl. 15, figs 15–16.
1990 *Dentacodina trilineatus* Wang; An and Zheng, pl. 15, fig. 17.
? 1990 *Dentacodina* aff. *D. dubia* (Rhodes, 1953); Armstrong, p. 72, pl. 20, figs 17–22 (multielement).
1992 *Dentacodina dubia* (Rhodes); Qian *in* Jin *et al.*, pl. 3, fig. 1.
v1996 *Dentacodina shaannanensis* Ding and Li; Wang and Aldridge, pl. 5, fig. 8.
v2002 *Dentacodina shaannanensis* Ding and Li; Aldridge and Wang, fig. 66H (copy of Wang and Aldridge 1996, pl. 5, fig. 8).

Diagnosis. Pa element with two or three laterally compressed, erect denticles; inner lip of basal cavity flared, without inner lateral process. M element erect or only very gently curved.

Material. Pa, 18; Pb, 97; Pc, 189; M, 223; Sa/Sb/Sd?, 112; Sc, 243; coniform, 544; plus additional material from TT samples.

Description. Pa element small, scaphate with just two or three broad, compressed denticles, the tallest of which is at one end, regarded to be anterior. Anterior edge of largest denticle sharp, angled a little posteriorly, with a straight or slightly concave

---

EXPLANATION OF PLATE 8

Figs 1–17. *Pseudobelodella spatha* (Zhou *et al.*, 1981). Yangpowan Member, Ningqiang Formation, Yushitan Section, Ningqiang County, Shaanxi, Sample Ningqiang 4. 1–2, 149715, P? element, lateral views. 3–4, 149716, S element, outer and inner lateral views. 5–6, 149717, S element, outer and inner lateral views. 7–9, 149718, P? element, lateral view, close-up of aboral margin to show wrinkle zone, lateral view. 10–11, 149719, S element, inner and outer lateral views. 12–13, 149720, S element, outer and inner lateral views. 14–15, 149721, S element, inner and outer lateral views. 16–17, 149722, S element, outer and inner lateral views.

Figs 18–31. *Pseudobelodella spatha* (Zhou *et al.*, 1981). Upper member, Xiushan Formation, Leijiatun Section, Shiqian County, Guizhou, Sample Shiqian 17. 18–19, 149723, P? element, lateral views. 20–21, 149724, S element, outer and inner lateral views. 22–23, 149725, S element, outer and inner lateral views. 24–25, 149726, S? element, inner and outer lateral views. 26–27, 149727, S element, outer and inner lateral views. 28–29, 149728, S element, outer and inner lateral views. 30–31, 149729, S element, inner and outer lateral views.

Fig. 32. *Pseudobelodella* sp. Shenxuanyi Member, Xuanhe Section, Guangyuan County, Sichuan, Sample Xuanhe 8. 149730, fused pair of S elements, outer lateral view, far element with alternating denticulation.

All figures ×120, except fig. 8, ×240.

PLATE 8

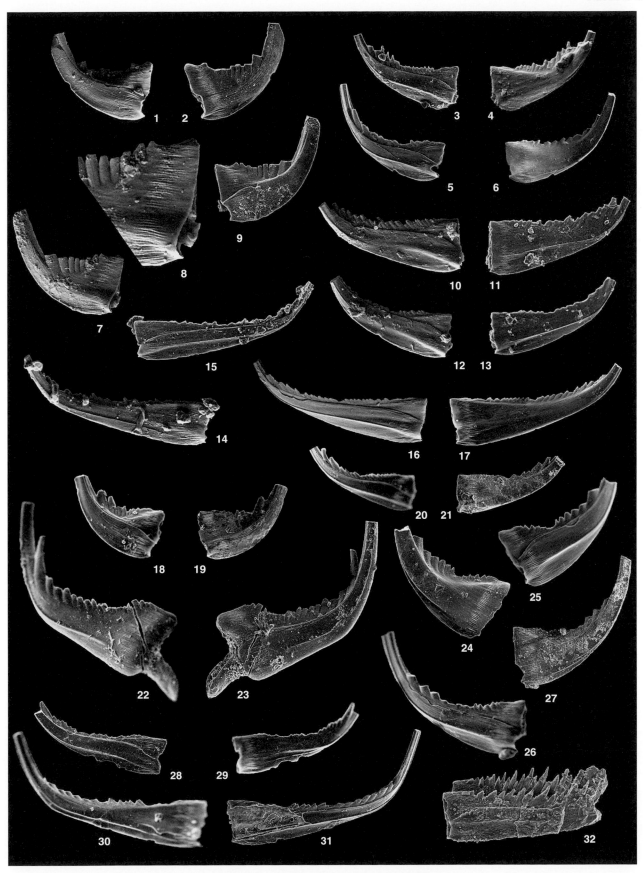

WANG and ALDRIDGE, *Pseudobelodella*

lateral profile. Posterior edge also sharp, straight or slightly concave. Denticles may be fused; all have smooth faces. Basal cavity wide and deep with an ovoid margin, extending under entire element. Lamellae edges visible within basal cavity (Pl. 9, fig. 3). White matter fills all denticles.

Pb element pastinate with tall, twisted cusp and triangular base. A small denticle may be developed on the inner lateral margin. Anterior and posterior processes very short; anterior process adenticulate, posterior process adenticulate or with a single erect, compressed denticle.

Pc element erect, with slightly twisted, posteriorly curved cusp. Posterior and anterior faces very gently convex, cross-section compressed and lenticular with sharp lateral edges; longitudinal microstriae faintly developed. Outline of aboral edge oval, with one margin developed into a short process that bears one or two erect denticles. White matter fills cusp. Basal cavity shallow, sometimes containing remains of lamellar basal body (Pl. 9, fig. 6).

M element makellate. Cusp stout, erect, with flaring posterior face towards base; lateral edges sharp, cross-section compressed and lenticular. Longitudinal microstriae may be well developed. Inner lateral margin of cusp extended aborally into a very short anticusp, adenticulate or with an incipient denticle. Outer lateral process very short, adenticulate or bearing one or two small, compressed denticles. Basal surface entirely excavated.

Sa element alate, symmetrical or very nearly symmetrical. Cusp stout, erect, maybe slightly twisted. Posterior process adenticulate to very short with one or two erect denticles, which may be only weakly joined to the cusp at its base; antero-lateral processes very short, adenticulate or with one or two conical denticles on each side. Basal cavity deeply excavated, flaring posteriorly; posterior denticles have separate basal cavities.

Sb element tertiopedate, strongly asymmetrical and quite variable. Cusp stout, erect, variably twisted, with an antero-lateral costa on each side, one costa extending into short lateral process bearing one to three denticles, the other extending downwards into a short process, adenticulate or with one or two conical

denticles. Posterior process very short with one or two denticles with ovoid transverse sections, commonly directed slightly to one side. Basal cavity deeply excavated.

Sc element. Cusp long, posteriorly recurved, plano-convex in section. Posterior process short with one to five compressed denticles. Costa on outer lateral face extends into very short rounded lateral process. Anterior face has a sharp or slightly costate lateral edge that extends into an adenticulate process, which may be extended as a long, thin anticusp (Pl. 9, fig. 22), commonly broken away. Basal cavity deeply excavated.

Sd? element alate, nearly symmetrical, with tall slightly twisted cusp with very well-developed lateral costae; these extend downwards into short lateral processes that may each bear a small, erect denticle. Posterior process very short with a single denticle that is directed posteriorly. Aboral margin of anterior face extended downwards into a small adenticulate lip, distinguishing this element from the Sa and Sb elements.

Coniform elements. Simple, mostly broad-based cones, either laterally (Pl. 9, figs 24–25) or antero-posteriorly (Pl. 9, figs 23, 26) compressed, oval or biconvex in midheight cross-section. Lower surface entirely and shallowly excavated; growth tip prominent in centre of cavity, edges of lamellae may be visible. Cavity outline circular, ovoid or asymmetrical. Sets of two or three cones weakly fused together or conjoined only by basal tissue are occasionally found (Pl. 9, fig. 27). As noted by previous workers (Jeppsson 1972; van den Boogaard 1990; Miller and Aldridge 1993), these discrete coniform elements appear to represent denticles from the processes of multidenticulate elements that have either broken away or were attached to the parent element only by basal tissue in life.

Remarks. Ding and Li (1985) illustrated three new species that might be elements of *Coryssognathus*: *Exochognathus ningqiangensis* (pl. 1, fig. 11), *Dentacodina ningqiangensis* (pl. 1, fig. 22) and *Dentacodina shaannanensis* (pl. 1, figs 24–25). The first two of these might equally be elements of *Distomodus*, so we have used the name *shaannanensis*

Figs 1–18, 20–21, 23–26. *Coryssognathus shaannanensis* (Ding and Li, 1985). Upper member, Xiushan Formation, Leijiatun Section, Shiqian County, Guizhou, Sample Shiqian 20. 1–3, 149731, Pa element, inner later and oral views and close-up to show lamellae edges within basal cavity. 4, 149732, Pa element, inner lateral view. 5, 149733, Pb element, lateral view. 6–7, 149734, Pc element, posterior and oral views. 8, 149735, M element, posterior view. 9, 149736, M element, posterior view. 10, 149737, M element, posterior view. 11, 149738, Sc element, outer lateral view. 12, 149739, Sb element, oblique posterior view. 13, 149740, Pc element, posterior view. 14–15, 149741, Sa element, lateral and oral views. 16, 149742, Sd? element, posterior view. 17, 149743, Sb element, posterior view. 18, 149744, Sb element, posterior view. 20–21, 149745, Sc element, inner and outer lateral views. 23, 149746, coniform element, posterior view. 24–25, 149747, coniform element, lateral views. 26, 149748, coniform element, posterior view.

Figs 19, 22. *Coryssognathus shaannanensis* (Ding and Li, 1985). Shenxuanyi Member, Xuanhe Section, Guangyuan County, Sichuan, Sample Xuanhe 11. 19, 149749, Sa element, lateral view. 22, 149750, Sc element, outer lateral view.

Fig. 27. *Coryssognathus shaannanensis* (Ding and Li, 1985). Shenxuanyi Member, Xuanhe Section, Guangyuan County, Sichuan, Sample Xuanhe 3, 149751, conjoined coniform elements, lateral view.

All figures ×100, except fig. 3, ×500.

PLATE 9

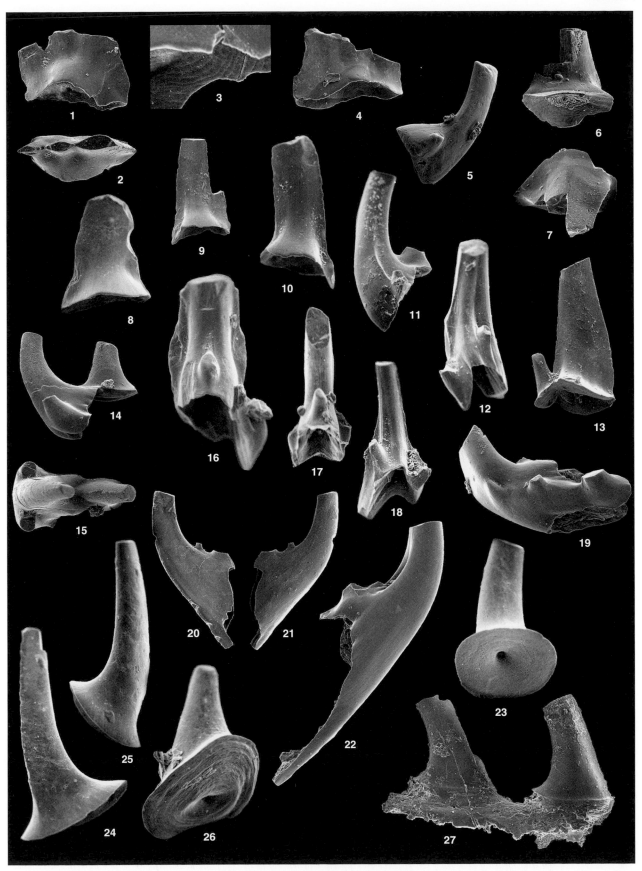

WANG and ALDRIDGE, *Coryssognathus*

for this species of *Coryssognathus*. The Chinese collections vary quite considerably; the population from Shiqian 20 comprises relatively poorly denticulated specimens with weakly developed costae, but similar specimens intergrade in other samples with specimens with stronger costae and more denticles, as exemplified by the holotype.

The specimens illustrated by Armstrong (1990) are very similar to the Chinese specimens and may come from the same species, or one that is very closely related. The scaphate Pa element was not found in the Greenland material, so cannot be compared. The specimen figured by Armstrong (1990) as the Pa (pl. 20, fig. 17) is a detached coniform (see above); his Pb (pl. 20, fig. 18) is probably a Pc; his M (pl. 20, fig. 19) is not an M element, and appears to be an Sc, in which case it is very different from that of *C. shaananensis* in having a diminutive cusp and large denticles on the posterior process.

Coryssognathus shaananensis is closely similar in many respects to *C. dubius* (Rhodes, 1953) (see van den Boogaard 1990; Miller and Aldridge 1993). The Pa element differs in the compressed, partly fused denticles, in the flaring of the inner lip of the basal cavity and in the lack of a secondary process. The M element is more erect in *C. shaananensis*, but the morphology of all other elements is almost identical.

Occurrence. Upper member, Xiushan Formation, Leijiatun section, Shiqian County, Guizhou; Yangpowan and Shenxuanyi members, Ningqiang Formation, Yushitan section, Ningqiang, Shaanxi; Shenxuanyi Member, Xuanhe section, Guangyuan, Sichuan.

Genus DISTOMODUS Branson and Branson, 1947

1947 *Distomodus* Branson and Branson, p. 553.
1964 *Hadrognathus* Walliser, p. 35.
1970 *Exochognathus* Pollock, Rexroad and Nicoll, p. 751.
1976 *Distomodus* Branson and Branson; Barrick and Klapper, p. 70 (multielement).
1977 *Johnognathus* Mashkova, p. 127.
1989 *Disciconus* Yu *in* Jin *et al.*, p. 101.

Type species. Distomodus kentuckyensis Branson and Branson, 1947.

Diagnosis. Emended after Bischoff (1986, p. 95): Pa element scaphate with four to six processes joining in the central region; oral surface with denticles, nodes and/or irregular ridges. Pb element tertiopedate with large cusp and large basal cavity; denticulate processes may develop platform ledges which may be smooth or covered with nodes and irregular ridges on the upper surface; posterior process may split into two processes. Pc element pastinate or bipennate with short or totally reduced processes; cusp

large and erect, basal cavity flaring widely. M element dolabrate with an anticusp that may bear denticles; cusp very large. Sc element bipennate to modified tertiopedate with denticulate inner antero-lateral and posterior processes and a short lip-like to long spur-like, downwardly directed, adenticulate outer projection. Sb element bipennate to modified dolabrate with short posterior process bearing few denticles, a strongly reduced, commonly denticulate anterolateral process, and a strongly reduced, adenticulate, downwardly directed projection similar to that of the Sc element. An Sd element is identifiable in some species, with well-developed posterior and (two) lateral processes and a short anterior lip. Sa element tertiopedate; processes denticulate, posterior process may develop platform ledges. Pa element without micro-ornament on upper surface; cusp and denticles of all other elements with very fine, longitudinal, subparallel striae.

Remarks. The holotype of the type species of *Disciconus*, *D. erlangshanensis* Yu, 1989 (Yu *in* Jin *et al.* 1989, p. 101, pl. 4, figs 2a, 2b, 9) is an element from the apparatus of *Distomodus*. Similar specimens were assigned by Bischoff (1986, pl. 7, figs 3, 7, 10, 11, 12) to the M position of his new species *Distomodus tridens*, but it is more likely that this type of element occupied a Pc position in the *Distomodus* apparatus (Wang and Aldridge 1998). The morphology of the Pa element is diagnostic for *Distomodus* species, and without knowing the nature of Pa specimens associated with the holotype of *D. erlangshanensis*, it is impossible to determine to which species it should belong. *Johnognathus* 'Pa elements' appear to represent broken or *in vivo* detached processes of *Distomodus* Pa and Sa elements (see Over and Chatterton, 1987, pl. 2, figs.13–16, 21–25).

Distomodus cathayensis sp. nov.
Plate 10, figures 1–17

?1983 *Distomodus kentuckyensis* Branson et Branson 1947; Zhou and Zhai, p. 273, pl. 65, figs 22a–b?, 23a–b (M element).
?1983 *Exochognathus brassfieldensis* (Branson et Branson, 1947); Zhou and Zhai, p. 275, pl. 65, fig. 27 (Sb element).
?1983 *Exochognathus caudatus* (Walliser, 1964); Zhou and Zhai, p. 275, pl. 65, fig. 29a–b (Sa element).
 1983 *Hadrognathus staurognathoides* Walliser, 1964; Zhou and Zhai, p. 277, pl. 66, fig. 4 (Pa element).
 1985 *Hadrognathus staurognathoides* Walliser; Yu, pl. 1, fig. 17 (Pa element).
p 1990 *Distomodus staurognathoides* (Walliser, 1964); Armstrong, p. 73, pl. 8, figs 6, 8–10 only [*non* fig. 7] (multielement).

1996 *Distomodus* sp. nov. Wang and Aldridge, pl. 3, fig. 4 (Pa element).

2002 *Distomodus* sp. nov. Aldridge and Wang, fig. 64D [copy of Wang and Aldridge 1996, pl. 3, fig. 4] (Pa element).

Derivation of name. From 'Cathaya', referring to its geographical occurrence in south China –*ensis*, L., suffix denoting place.

Holotype. Specimen NIGPAS 117112, Pl. 10, figs 1–2; Pa element.

Type locality and horizon. Lower member, Xiushan Formation, Leijiatun section, Shiqian County, Guizhou (sample TT 752b).

Diagnosis. Distomodus species characterized by a Pa element with five processes, each bearing a median ridge; these ridges do not converge to a single point but meet at three points that are widely separated. Other elements lack development of platform ledges.

Material. Pa, 38; Pb, 71; Pc, 59; M, 79; Sa, 24; Sb, 96; Sc, 119; Sd, 27; plus additional material from TT samples.

Description. Pa element large, pastiniscaphate, comprising five processes; it is difficult to determine which are primary and which secondary. Processes do not radiate from a single point but meet at three distinct triple junctions, which are joined by two low ridges; these ridges form a straight line or a highly obtuse angle. One process, although almost always broken, distinctly longer than the others, slightly inclined, and bears a carina consisting of small, low, discrete denticles, which increase in height distally; one or two particularly large denticles situated close to end of process. The other four processes of subequal length, straight or gently curved; each bears an axial carina on upper surface formed by low denticles, which may be discrete but more commonly joined by ridges. Entire aboral surface deeply excavated; basal cavity deepest beneath region in which processes join; a lamellar basal body often preserved. White matter restricted to area of raised ornament on oral surface. Specimens rarely complete; fragments may resemble pieces of the Pa element of *Distomodus staurognathoides*.

Pb element tertiopedate with large, slightly twisted cusp. Three prominent costae on cusp give rise to short posterior, inner lateral and outer lateral processes; outer lateral process is the longest and commonly bears up to four discrete, flattened denticles that are triangular in posterior view. Inner lateral process may also bear one or two denticles that may be fused into a low ridge; posterior process normally adenticulate. Basal cavity triangular, extending to ends of processes.

Pc element erect, pastinate, with slightly twisted cusp. Cusp with a lateral costa on each side; these costae not developed into true processes, but may, in large specimens, give rise to one or two denticles situated on oral surface of flaring basal cavity. Basal cavity subcircular, sometimes with an irregular margin, flaring widely and shallowly.

M element makellate, with large robust cusp that is curved slightly posteriorly. Cross-section of cusp biconvex, with sharp lateral edges. Convex lateral margin extended downwards to form an adenticulate anticusp. The concave edge gives rise to a very short process that bears a single denticle or a short ridge of fused denticles. Basal cavity deeply excavated, extending as a narrow groove beneath anticusp.

Sa element tertiopedate, subsymmetrical, with stout cusp, two lateral processes and a posterior process. The posteriorly curved cusp bears three prominent costae, which merge downwards with processes, which are of subequal length. Each lateral process bears two or three small denticles, which may become almost fused into a low ridge. Posterior process also bears two or three small, discrete denticles. Base deeply excavated, with cavity continuing as a narrow groove beneath processes.

Sb element asymmetrical, tertiopedate, dominated by tall, posteriorly curved, laterally twisted cusp. Base of cusp laterally and posteriorly expanded to form a triangular outline. Posterior edge prominent, extended into a very short posterior process, which bears one or two small denticles. Outer-lateral costa prominent, strongly twisted, merging with a lateral process which bears four or five partly fused small denticles. Inner-lateral costa relatively weak, extending into a process that bears discrete denticles which may increase in size distally. Cross-section of the cusp subtriangular. Basal cavity deeply excavated, continuing beneath all processes.

Sc element dolabrate. Cusp large, robust and curved posteriorly. Two sharp costae on cusp give rise to an antero-lateral process and a posterior process. Antero-lateral process extended downwards and inwards, adenticulate or bearing up to five discrete denticles. Posterior process straight, with one or two discrete small denticles, which may become fully fused. Outer lateral face of cusp rounded into a broad costa, which extends aborally into a short, adenticulate lip. Base deeply excavated, cavity continuing as a narrow groove beneath processes.

Sd element subsymmetrical, modified quadriramate. Cusp prominent and robust, slightly twisted with sharp lateral edges that are developed into short alae and extend downwards into short lateral processes that bear up to three discrete denticles. Posterior face of cusp bears a costa that extends into a short posterior process bearing up to four small denticles; anterior face rounded and extending downwards into an aborally projecting lip. Base deeply excavated, cavity extending as a groove under processes.

Remarks. Specimens of this new species have been assigned to *Distomodus staurognathoides* in previous works on the conodonts of south China (Zhou *et al.* 1981; Zhou and Zhai 1983; Liu *et al.* 1993), but they are clearly distinguishable by the pattern of radiation of the processes on the Pa element. In *D. staurognathoides*, all the processes on the Pa element radiate from a single central point and generally have irregular ornamentation on the upper surface. The specimens described and figured as *Distomodus staurognathoides* by Armstrong (1990, pp. 73–76) from Greenland include a Pa element with the processes offset in the same manner as shown by those from China; the Greenland specimens from the same sample as the illustrated Pa are here all considered to belong to *D. cathayensis*.

It is possible that the specimen illustrated as *Erisomodus shiqianensis* sp. nov. by Zhou *et al.* (1981, p. 131, pl. 1, figs 11–12) and refigured by Zhou and Zhai (1983, pl. 66, figs 17 a–b) is an Sa element of *D. cathayensis*; if this is the case, then the species name *shiqianensis* would have priority. However, there is no evidence for the nature of the Pa element associated with this specimen, so this assignment cannot be made.

Occurrence. Lower member, Xiushan Formation, Leijiatun section, Shiqian County, Guizhou; also found at Zheng'an, north Guizhou.

Distomodus cf. *combinatus* Bischoff, 1986
Plate 10, figure 18

cf. 1986 *Distomodus combinatus* Bischoff, p. 102, pl. 7, figs 19–26, pl. 8, figs 1–7, 12, 17–21 [*non* pl. 8, figs 13–14, probably M elements of *D. calcar* Bischoff] (multielement).

Material. One broken Pa element and an associated Sa element only.

Description. Pa element pastiniscaphate, broken, with just two processes preserved. Angle between processes about 60°. Longer process with axial ridge that extends almost to tip; unornamented proximally but with four pairs of nodes distally that are joined across the axial ridge by transverse ridges. Shorter process tapering distally to sharp point; bears an irregular, nodose axial ridge and three or four marginal nodes distally; about a third of the distance along its length a bifurcating ridge leads to a possible broken secondary process on the far side from the longer process.

Sa element tertiopedate with curved cusp supporting two lateral processes and a posterior process. Three costae extend along the cusp in line with the processes. Posterior process bears two laterally compressed denticles and is of similar length to the lateral processes. Each lateral process bears about four small laterally, compressed, fused denticles and is directed downwards.

Remarks. The angle between the processes of the Pa element and the ornamentation of the oral surface are comparable with *D. combinatus*, but there are a number of differences, including the unornamented proximal area on the longer process in our specimen. The bifurcating processes on *D. combinatus* have their joint origination close to the basal cavity tip, whereas the bifurcation occurs more distally in our specimen; the bifurcating processes are also relatively longer. The morphology of *Distomodus* Pa elements is variable within species, but these differences are too great for us to refer our broken specimen confidently to *D. combinatus*.

Occurrence. Xiangshuyuan Formation, Leijiatun Section, Shiqian County, Guizhou.

Family BALOGNATHIDAE Hass, 1959

Genus APSIDOGNATHUS Walliser, 1964

　　1964 *Apsidognathus* Walliser, p. 29.
　　1964 *Astrognathus* Walliser, p. 30.
　　1981 *Kailidontus* Zhou et al., p. 133.
　? 1981 *Nericodina* Zhou et al., p. 135.
　　1983 *Neopygodus* Zhou and Zhai, p. 284.
　? 1983 *Parapygodus* Zhou and Zhai, p. 291.
　　1986 *Pseudopygodus* Bischoff, p. 241.

Type species. Apsidognathus tuberculatus Walliser, 1964, p. 29.

Figs 1–3, 5–6, 9, 12, 14–15. *Distomodus cathayensis* sp. nov. Lower member, Xiushan Formation, Leijiatun Section, Shiqian County, Guizhou, Sample TT 752b. 1–2, 117112, Pa element, oral and aboral views (holotype). 3, 149752, Pa element, oral view. 5, 149753, Pb element, posterior view. 6, 149754, Pb element, posterior view. 9, 149755, M element posterior view. 12, 149756, Pc element, posterior view. 14, 149757, Sb element, posterior view. 15, 149758, Sb element, posterior view.

Figs 7–8, 10, 13, 17. *Distomodus cathayensis* sp. nov. Lower member, Xiushan Formation, Leijiatun Section, Shiqian County, Guizhou, Sample TT 752a. 7–8, 149759, Pc element, aboral and lateral views. 10, 149760, Sa element, posterior view. 13, 149761, Sd element, posterior view. 17, 149762, Sc element, posterior view.

Figs 4, 11, 16. *Distomodus cathayensis* sp. nov. Lower member, Xiushan Formation, Leijiatun Section, Shiqian County, Guizhou, Sample Shiqian 15. 4, 149763, Pa element, oral view. 11, 149764, M element, posterior view. 16, 149765, Sc element, outer lateral view.

Fig. 18. *Distomodus* cf. *combinatus* Bischoff, 1986. Xiangshuyuan Formation, Leijiatun Section, Shiqian County, Guizhou, Sample TT 813. 149766, Pa element, oral view.

All figures ×40.

PLATE 10

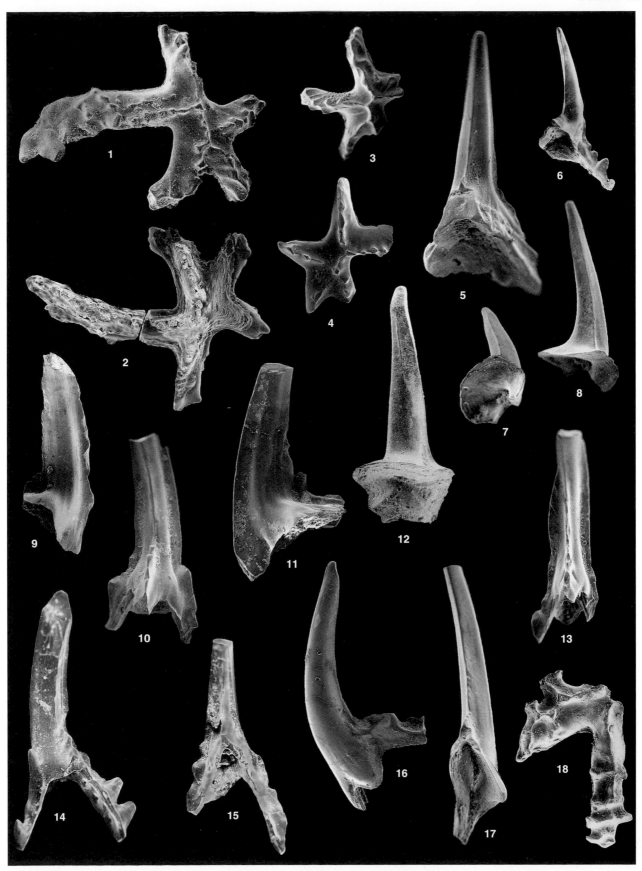

WANG and ALDRIDGE, *Distomodus*

Diagnosis. See Armstrong (1990, p. 41). The inclusion of the coniform element is equivocal.

Remarks. Sweet (1988) referred *Apsidognathus* to the family Pterospathodontidae, but the relationships of the genus are still obscure. There are no similar apparatuses known, and the ancestry of the genus is cryptic. There is a broad comparability of the compressed scaphate elements to the pygodiform element of *Anticostiodus* Zhang and Barnes, 2000 (see Aldridge 2002, p. 22, pl. 1, figs 6–7, 10–11), but the other elements are dissimilar, and *Anticostiodus* itself is an unusual genus with unknown relationships. According to Over and Chatterton (1987, p. 22, pl. 3, figs 11–12, 15–17), the type species of their new genus *Astrolecignathus*, *A. milleri* Over and Chatterton, also possessed a compressed scaphate element, but the complete apparatus is unknown.

The apparatus of *Apsidognathus* is also still enigmatic. Walliser (1972, p. 76) partially reconstructed the apparatus by including the form species *Pygodus lyra* Walliser, 1964, along with *A. tuberculatus*. Aldridge (1974) suggested that the form species *Ambalodus galerus* Walliser, 1964, might also be included, and *Astrognathus tetractis* Walliser, 1964, is also assigned to the apparatus, as suggested by Klapper (*in* Robison 1981, p. 136). Mabillard and Aldridge (1983) included compressed lenticular and conical elements in the apparatus of *A. ruginosus* Mabillard and Aldridge, and Armstrong (1990, p. 41) considered the apparatus of *Apsidognathus* to consist of seven element types: platform, lenticular, ambalodontan, astrognathodontan, lyriform, compressed and coniform. Armstrong (1990) suggested that the astrognathodontan (stelliscaphate) element could occupy either the Sa or Sd position in the apparatus; if the former is correct, the lyriform might be the Sb element, and the compressed element could be the Sc element. If the astrognathodontan element occupied the Sd position, it would then be possible that the lyriform occupied both Sa and Sb positions. This possibility is strengthened by the observable variation from symmetrical to asymmetrical specimens of the lyriform element (see below). However, the lyriform element might alternatively be an M, and our increased knowledge of prioniodontid apparatus templates indicates that Pc or Pc + Pd positions might also be present in the *Apsidognathus* apparatus. It is possible that the ornate elements fill up to five P and M positions and that the S elements are reduced to coniform morphology or are absent. Over and Chatterton (1987) suggested something similar, with the platform, compressed, astrognathodontan and ambalodontan elements assigned to Pa$_1$, Pa$_2$, Pb$_1$ and Pb$_2$ positions, respectively, and the lyriform identified as an S element. Within the P elements, it may be that the platform element is the Pa, and the ambalodontan is the Pb, but comparison with the apparatus of *Promissum*

(see Aldridge *et al.* 1995) suggests the alternative possibility that the Pa and Pb positions are both occupied by pastiniscaphate elements and that the ambalodontan occupies the Pc position. The discovery of a bedding plane assemblage is probably necessary before the apparatus of this genus can be understood.

Over and Chatterton (1987) suggested that *Tuxekania* Savage, 1985, is a junior synonym of *Apsidognathus*, but the platform element of the type species, *T. barbarajeanae* Savage, 1985, has a different style of basal cavity from that of *A. tuberculatus*. Given that the apparatus of *Tuxekania* is currently unknown, it is best retained as a separate genus.

The type species of the genus *Kailidontus*, *K. typicus* Zhou *et al.* (1981), is very similar to *Astrognathus* Walliser 1964. Walliser's type species, *A. tetractis*, has become widely regarded as a component of the multielement species *Apsidognathus tuberculatus* (e.g. Uyeno and Barnes 1981; Aldridge 1985; Bischoff 1986; Armstrong 1990). Specimens referred to *Kailidontus* almost certainly belonged to a similar apparatus and, as indicated by Fordham (1991, p. 58), the genus should be considered a subjective junior synonym of *Apsidognathus*. Similarly, Zhou and Zhai (1983) erected *Neopygodus* to accommodate specimens such as those originally placed in *Pygodus lyra* Walliser, 1964, but subsequently considered to be distinct from the Ordovician genus *Pygodus* Lamont and Lindström (e.g. Aldridge 1972). The two species Zhou and Zhai (1983) assigned to *Neopygodus*, *N. lyra* and the type species, *N. kailiensis*, are similar to each other and probably conspecific. Specimens of *P. lyra* have, however, long been regarded as forming part of the apparatus of *Apsidognathus* (e.g. Walliser 1964, 1972). *Neopygodus* is, therefore, a subjective junior synonym of *Apsidognathus*. Further, the genus *Parapygodus* was erected by Zhou and Zhai (1983) to accommodate lenticular specimens such as those originally placed in *Pygodus? lenticularis* Walliser, 1964. Specimens of *P? lenticularis* have also been widely regarded as forming part of the apparatus of *Apsidognathus* (e.g. Uyeno and Barnes 1981, 1983; Aldridge 1985; Armstrong 1990), but Bischoff (1986) placed similar specimens in his genera *Pseudopygodus* and *Pyrsognathus*. Fordham (1991, p. 33) proposed a bimembrate apparatus for *Parapygodus*, with *Pseudopygodus* implicitly included as a subjective junior synonym. The status of *Parapygodus* remains uncertain, although we consider the evidence to be strong for its inclusion as a junior synonym of *Apsidognathus* and have included elements of this type in our reconstructions of *Apsidognathus* apparatuses.

Compressed pygodiform elements are also possible members of the apparatus of *Tuberocostadontus*, and the assignment of particular specimens to *Apsidognathus* and *Tuberocostadontus* is currently equivocal. Here, we retain elements with a relatively straight oral margin in *Apsido-*

gnathus, where we also tentatively include some associated specimens with a more triangular outline and higher inner face. Other compressed specimens with a triangular outline are more closely associated with elements referred to *Tuberocostadontus*.

All the elements of the *Apsidognathus* apparatus show a very plastic morphology, and there is considerable variation in ornamentation of specimens in any large collection. This intraspecific variability makes the delimitation of species very difficult. The greatest morphological disparity is shown by the platform and lyriform elements, and the variability of these in the collections we have studied from China indicates that at least four species are present. Although we have attempted to differentiate these, some elements in small collections and in collections where more than one species occurs are currently impossible to assign unequivocally at species level.

Apsidognathus aulacis Zhou, Zhai and Xian, 1981
Plate 11, figures 1–24

*1981 *Apsidognathus aulacis* Zhou et al., p. 130, pl. 1, figs 1–2 (platform element).

?1981 *Kailidontus typicus* Zhou et al., p. 133, pl. 1, figs 24–27 (astrognathodontan element).

?1981 *Nericodina cricostata* Zhou et al., p. 135, pl. 1, figs 42–43 (compressed element).

1983 *Apsidognathus aulacis* Zhou, Zhai et Xian, 1981; Zhou and Zhai, p. 269. pl. 65, fig. 7a–b (platform element).

?1983 '*Ambalodus*' *galerus* Walliser, 1964; Zhou and Zhai, p. 268, pl. 65, figs 2a–b, 3a–b (ambalodontan element).

?1983 *Kailidontus typicus* Zhou, Zhai et Xian, 1981; Zhou and Zhai, p. 278, pl. 66, fig. 6a–b (astrognathodontan element).

?1983 *Neopygodus kailiensis* Zhou and Zhai, p. 284, pl. 67, fig. 22a–b (lyriform element).

?1983 *Neopygodus lyra* (Walliser, 1964); Zhou and Zhai, p. 285, pl. 67, fig. 21a–b (lyriform element).

?1983 *Nericodina cricostata* Zhou, Zhai et Xian, 1981; Zhou and Zhai, p. 286, pl. 67, fig. 4a–b (compressed element).

?1987 *Kailidontus macrodentatus* An, p. 198, pl. 33, fig. 11 (astrognathodontan element).

?1987 *Neopygodus kailiensis* Zhou and Zhai 1983; An, p. 199, pl. 33, fig. 14 (lyriform element).

?1987 *Neopygodus lyra* (Walliser) 1964; An, p. 199, pl. 33, fig. 13 (lyriform element).

?1987 Gen. & sp. indet. F An, pl. 31, figs 19–21 (ambalodontan element).

?1992 *Ambalodus galerus* Walliser, 1964; Qian *in* Jin *et al.*, p. 58, pl. 2, fig. 2 (ambalodontan element).

1992 *Apsidognathus aulacis* Zhou, Zhai and Xian, 1981; Qian *in* Jin *et al.*, p. 58, pl. 2, fig. 6 (platform element).

?1992 *Neopygodus lyra* (Walliser, 1964): Qian *in* Jin *et al.*, p. 60, pl. 2, fig. 11 (lyriform element).

?1992 *Astrognathus tetractis* Walliser; Qian *in* Jin *et al.*, pl. 3, fig. 3 (astrognathodontan element).

v 1996 *Apsidognathus aulacis* Zhou et al.; Wang and Aldridge, pl. 4, figs 9–10 (platform element).

v 2002 *Apsidognathus aulacis* Zhou et al.; Aldridge and Wang, figs 65I–J. [copy of Wang and Aldridge 1996, pl. 4, figs 9–10] (platform element).

Diagnosis. Platform element characteristic, with a subquadrate outline to the platform and variable nodose ornament with a concentration of the nodes at the platform margins; ambalodontan element with a widely flaring basal cavity; lyriform element with a weak axial ridge; stelliscaphate astrognathodontan element with four processes not all of equal length.

Material. Platform, 212; ambalodontan, 124; lyriform, 63; astrognathodontan, 17; compressed, 73; plus additional material from TT samples.

Description. Platform element broad, scaphate. Free blade short, continuing as a curved carina across the platform; carina reduces in height posteriorly, terminating short of posterior tip in mature specimens (Pl. 11, figs 6–7). Inner platform narrower than outer platform, with a strong shoulder anteriorly and a straight margin for most of its length. Outer platform with two lobes, variably prominent, with the posterior one commonly the stronger. Oral surface of platform variably ornamented with nodes, which are commonly prominent along the margins, particularly on straight edge of inner platform; nodes show a vague to more pronounced concentric lamination and may be connected by low, inconspicuous ridges radiating from apex. A narrow smooth area may occur alongside carina on inner side; proximal portion of outer platform also sometimes smooth. Small, presumably juvenile specimens may have an entirely smooth platform except at margin of inner platform (Pl. 11, fig. 19). Tips of nodes on carina and on platform show polygonal microstructure (Pl. 11, figs 3–4). White matter fills denticles on free blade and carina, with shallow roots.

Ambalodontan element arched, anguliscaphate, with prominent cusp. Anterior and posterior processes form angle of 140–170° in oral view; denticles on both processes triangular, variably spaced, sometimes fused proximally. Cavity flares on both sides; on inner side the flare does not extend to anterior tip. Inner lobe of flare commonly smooth, sometimes with a single medial row of nodes and, in larger specimens, may bear scattered small nodes; outer lobe narrower, smooth. White matter fills cusp and denticles, forming an almost continuous block.

Lyriform element arched, with a short free blade comprising one to three fused denticles. Platform expands posteriorly, with lateral margins bearing ridges formed of fused denticles. The free blade may align with one of these ridges (Pl. 11, fig. 20), or may lie symmetrically between them (Pl. 11, fig. 10); there is a small shoulder on one or both sides at junction of blade and platform. Platform thin and smooth between the lateral ridges, with a low, continuous medial ridge that sometimes extends as a small posterior lip beyond rest of platform; posterior end, thereby, sometimes concave, or sometimes shows two concave indentations. Beneath the lateral ridges thin lateral walls angle inwards aborally to enclose broad basal cavity, deepest beneath blade/platform junction. White matter fills the fused denticles of the free blade and the lateral ridges.

Astrognathodontan element stelliscaphate, with strongly arched blade and shorter, less strongly arched processes. Blade almost symmetrical, bearing low denticles that are almost entirely fused. Processes in the form of lobes with pointed terminations, oriented perpendicularly on either side of blade, commonly bearing a medial ridge and sometimes with low, irregular nodes. Basal cavity wide and deep below apex of unit, extending beneath all processes. White matter forming a continuous block along the fused denticles of the blade.

Compressed element 1 strongly compressed so that inner and outer faces almost in contact. Denticle row fused to low curved ridge with no conspicuous cusp. Outer face rounded triangular to ovoid in shape and variably ornamented with nodes and rugae, but with curved low rows of rugae parallel or subparallel to the lower margin dominant. Inner face much shorter, smooth with straight basal edge. Basal cavity narrow and deep. White matter fills the denticular ridge.

?Compressed element 2 with denticle row strongly arched with prominent cusp; other denticles fused to a low, serrated ridge. Outer face rounded triangular in shape with medial row of low nodes extending to a basal lip and other irregularly scattered nodes. Inner face much shorter, smooth, with a gently convex basal edge. Basal cavity narrow and deep. White matter fills cusp and denticular ridge.

Remarks. This species was named on the basis of a single well-preserved specimen of the platform element by Zhou *et al.* (1981). They considered that it differed from the corresponding element of *A. tuberculatus* in having a groove near the carina and in the lack of concentric nodose striations. However, some specimens of *A. tuberculatus* illustrated by Walliser (1964, pl. 12, fig. 16, p. 13, fig. 2) have a similar groove and not all possess concentric rows of nodes. Our larger collections reveal that several specimens of *A. aulacis* lack the groove, and the subquadrate outline of the platform and the concentration of nodes at the platform margin are more consistent characters. The lyriform elements are morphologically close to those figured by Walliser (1964) as *Pygodus lyra* (especially pl. 12, figs 8–10, 13–14), although all the specimens he illustrated lack a medial ridge. A specimen from Gullet Quarry, England, with a more strongly developed, nodose medial ridge was illustrated by Aldridge (1975, pl. 1, fig. 2; 1985, pl. 3.2, fig. 24) as the lyriform element of *A. tuberculatus*.

We differ from Armstrong (1990, p. 42) in considering the more strongly arched processes of the astrognathodontan element to form the blade; we cannot determine whether these processes are lateral or anterior–posterior. The astrognathododontan element figured by Walliser (1964, pl. 14, fig. 1–2) as *Astrognathus tetractis*, and subsequently considered to be an element of the apparatus of *Apsidognathus tuberculatus*, is more equantly cruciform in shape with all the processes denticulate and of similar height.

For the first time, we tentatively differentiate two compressed elements in an *Apsidognathus* apparatus. These may reflect two element positions, or a considerable variability of morphology within one element position. The possibility of two separate positions is also suggested for the apparatus of *A. ruginosus scutatus*. However, the second compressed element might alternatively belong to the apparatus of *Tuberocostadontus* (see below).

EXPLANATION OF PLATE 11

Figs 1–14, 20. *Apsidognathus aulacis* Zhou *et al.*, 1981. Shenxuanyi Member, Xuanhe Section, Guangyuan County, Sichuan, Sample Xuanhe 2. 1–4, 149767, platform element, oral and lateral views, close-ups to show reticulate micro-ornament on platform nodes and on denticles on the carina. 5, 149768, ambalodontan element, upper view. 6–7, 149769, platform element, oral and lateral views. 8, 149770, ambalodontan element, lateral view. 9–10, 149771, lyriform element, lateral and oral views. 11–12, 149772, astrognathodontan element, lateral? and oblique views. 13–14, 149773, compressed element 1, outer and inner lateral views. 20, 149774, lyriform element, oral view.

Figs 15–18. *Apsidognathus aulacis* Zhou *et al.*, 1981. Shenxuanyi Member, Xuanhe Section, Guangyuan County, Sichuan, Sample Xuanhe 1. 15–16, 149775, compressed element 1, outer and inner lateral views. 17–18, 149776, ?compressed element 2, inner and outer lateral views.

Figs 19, 21–23. *Apsidognathus aulacis* Zhou *et al.*, 1981. Shenxuanyi Member, Xuanhe Section, Guangyuan County, Sichuan, Sample TT 488. 19, 149777, platform element, oral view. 21, 149778, astrognathodontan element, oral view. 22–23, 149779, ambalodontan element, lateral and oral views.

Fig. 24. *Apsidognathus aulacis* Zhou *et al.*, 1981. Shenxuanyi Member, Xuanhe Section, Guangyuan County, Sichuan, Sample TT 613. 149780, platform element, oral view.

All figures ×60, except fig. 3, ×200, fig. 4, ×600.

PLATE 11

WANG and ALDRIDGE, *Apsidognathus*

The elements assigned by Zhou *et al.* (1981) to *K. typicus* and *Nericodina cricostata* are probably elements of *Apsidognathus*, but without knowing the nature of associated elements we cannot be sure of the correct specific assignment.

Occurrence. Upper member, Xiushan Formation, Leijiatun section, Shiqian County, Guizhou; Yangpowan and Shenxuanyi members, Ningqiang Formation, Yushitan section, Ningqiang, Shaanxi; Shenxuanyi Member, Xuanhe section, Guangyuan, Sichuan.

Apsidognathus ruginosus Mabillard and Aldridge, 1983

p 1983 *Apsidognathus ruginosus* Mabillard and Aldridge, p. 32, pl. 1, figs 1–2, 5–9, 14, ?figs 3–4, 10–11 [*non* figs 12–13] (multielement).

Diagnosis. See Mabillard and Aldridge (1983, p. 32).

Remarks. Several specimens in the Chinese collections conform to the diagnosis. However, there is considerably more variability in some elements than described by Mabillard and Aldridge (1983), and a few consistent differences from the British material. We consider that the Chinese population may be a genuine geographical variant within this species, so we have afforded it subspecific status.

Some specimens referred to *A. tuberculatus* in the literature show characteristics of *A. ruginosus* (e.g. Over and Chatterton 1987, pl. 4, figs 7 (platform), 10 (astrognathodontan), 21 (lyriform)), suggesting that this species might be more widely distributed than previously appreciated.

Apsidognathus ruginosus scutatus subsp. nov.
Plate 12, figures 1–18

?1983 *Parapygodus hemiorbicularis* Zhou and Zhai, p. 292, pl. 67, fig. 23a–b (compressed element 1).

?1983 *Parapygodus ovalis* Zhou and Zhai, p. 292, pl. 67, fig. 24a–b (compressed element 2).
?1985 *Parapygodus triangularis* Ding and Li, p. 17, pl. 1, figs 1–8 (compressed element 1).
1992 *Apsidognathus tuberculatus* Walliser, 1964; Qian *in* Jin *et al.*, p. 59, pl. 2, fig. 7 (platform element).
1992 *Parapygodus triangularis* Ding; Qian *in* Jin *et al.*, pl. 2, fig. 10 (compressed element 1).

Derivation of name. Latin, *scutatus*, armed with a shield; with reference to the shield shape of the characteristic elongate compressed element.

Holotype. Specimen NIGPAS 149785, Pl. 12, figs 12–14; elongate compressed element.

Type locality and horizon. Xuanhe section, Guangyuan County, Sichuan; Shenxuanyi Member, sample Xuanhe 3.

Diagnosis. Subspecies of *A. ruginosus* with a distinctive compressed element, elongate and shield-shaped with prominent concentric rugae. Lyriform element with prominent transverse ridges that may be separated by a medial trough or may meet across the axis of the element.

Material. Platform, 42; ambalodontan, 13; lyriform, 9; astrognathodontan, 2; compressed, 24; plus additional material from TT samples.

Description. Platform element broad, scaphate, morphologically variable, but with a consistent pattern of concentric rugose ornament on platform. Short free blade continues across platform as a carina, which is curved or deflected sharply inwards at the apex. Free blade and carina with fused denticles; denticles on free blade and anterior part of carina narrow and pointed, those on posterior part of carina characteristically broad, becoming a set of transverse ridges in mature specimens (Pl. 12, fig. 1). Outer platform more broadly flared than inner platform, with two weak lobes coincident with faint radiating rows of nodes that arise from separate points on the carina. Inner platform subquadrate in outline, often with a strongly developed nodose

---

EXPLANATION OF PLATE 12

Figs 1–5, 10–14. *Apsidognathus ruginosus scutatus* subsp. nov. Shenxuanyi Member, Xuanhe Section, Guangyuan County, Sichuan, Sample Xuanhe 3. 1, 149781, platform element, oral view. 2–4, 149782, lyriform element, lateral, oral and aboral views. 5, 149783, astrognathodontan element, oral view. 10–11, 149784, ?compressed element 2, outer and inner lateral views. 12–14, 149785, compressed element 1, oral, aboral and lateral views (holotype).

Figs 6–8, 15–18. *Apsidognathus ruginosus scutatus* subsp. nov. Daluzhai Formation, Hunggexi Section, Daguan County, Yunnan, Sample TT 1169. 6. 149786, ambalodontan element, lateral view. 7, 149787, platform element, oral view. 8, 149788, lyriform element, oral view. 15–16, 149789, ?compressed element 2, outer and inner lateral views. 17, 149790, astrognathodontan element, oral view. 18, 149791, astrognathodontan element, oblique oral view.

Fig. 9. *Apsidognathus ruginosus scutatus* subsp. nov. Shenxuanyi Member, Xuanhe Section, Guangyuan County, Sichuan, Sample Xuanhe 4. 149792, platform element, oral view.

All figures ×60.

PLATE 12

WANG and ALDRIDGE, *Apsidognathus*

antero-lateral ridge directed at about 30° to carina and a less strongly developed postero-lateral row of nodes; the rows of nodes converge on a single area of the carina, but may be separated from it by a short perpendicular ridge (Pl. 12, fig. 7). Platform surface with ornament of concentric low ridges parallel to platform margin; trough between carina and inner antero-lateral ridge may be smooth. White matter fills nodes on free blade, carina and inner antero-lateral ridge.

Ambalodontan element anguliscaphate. Arched, pyramidal unit with prominent cusp and widely flared cavity; posterior process generally much shorter than anterior process. On small specimens, surfaces of cavity lips smooth; on larger specimens, there may be rows of nodes parallel to the margins. Base deeply excavated. White matter fills cusp and all denticles.

Lyriform element arched, bilaterally symmetrical or nearly so. Anterior free blade short and, in larger specimens, broad, bearing nodes or transverse rugae. Platform commonly with medial trough, either side ornamented by ridges perpendicular to the axis; these may join across the trough and also create a serrated margin to the edge of the platform in oral view. Posterior end of platform broken away on all specimens, so shape of termination unknown. Lateral walls smooth beneath serrated margin directed a little inwards aborally and enclosing a deep cavity.

Astrognathodontan element stelliscaphate; anterior and posterior processes forming a strongly arched ridge, which is straight in oral view. Distal end of anterior process may be gently flexed. Arch bears fused denticles without a prominent cusp. Lateral process on each side of arch, directed perpendicularly, one twice the length of the other; both with a straight, horizontal aboral edge and both with a row of low, fused denticles that decrease in height distally; on some specimens, longer process bears an ornament of irregular low nodes, rather than a medial row. White matter fills denticles of the ridge and of processes as a continuous block. Basal cavity deep, extending as a wide groove to the tips of all processes.

Compressed element 1 very compressed, with the outer lateral face extended vertically downwards to several times the length of the inner lateral face. Denticles fused into a straight or undulating serrate ridge, without a prominent cusp. Inner process smooth with a convex basal edge. Outer process platform-like, narrowing distally to give the unit a shield shape, with a rounded distal termination; ornament consists of a short vertical ridge beneath the midpoint of the denticle row that is surrounded by concentric, well-developed ridges, horseshoe-shaped proximally and becoming transverse distally. Edges of growth lamellae clearly marked on underside of outer process. White matter forms a continuous thin block along the denticle ridge.

?Compressed element 2 very compressed, with inner lateral face less strongly developed than outer lateral face. Denticle row arched with a prominent cusp, directed gently posteriorly and sometimes also inwards. Anterior process steep, adenticulate; posterior process with fused low denticles forming a serrate ridge. Inner lateral face of cusp with a faint medial ridge; outer lateral face with 1–3 faint vertical ridges. Inner lateral process near vertical, short, smooth, with a straight or gently undulose basal edge. Outer lateral process near vertical, with a variably developed lip extending downwards towards the posterior end; ornamented by concentric nodes or low concentric ridges. White matter fills cusp deeply and extends along the denticles of the posterior process.

Remarks. The elements show a strong rugose ornament and, in this respect, closely resemble the nominate subspecies from Wales (Mabillard and Aldridge 1983). However, while some specimens of the lyriform element are closely comparable with the specimen illustrated by Mabillard and Aldridge (1983, pl. 1, fig. 9), others lack the medial trough and the ridges are continuous across the element (Pl. 12, figs 3–4). Additionally, although two distinct lenticular morphologies are present, as also apparent in the British collection (Mabillard and Aldridge 1983, pl. 1, figs 3–6), the squatter element sometimes has a more pronounced cusp in the Chinese material (Pl. 12,

---

EXPLANATION OF PLATE 13

Fig. 1. *Apsidognathus tuberculatus* Walliser, 1964. Shenxuanyi Member, Xuanhe Section, Guangyuan County, Sichuan, Sample Xuanhe 2. 149793, platform element, oral view.

Figs 2–4. *Apsidognathus* aff. *arcticus* Armstrong, 1990. Ningqiang Formation, Dazhuba Section, Ninqiang County, Shaanxi, Sample TT 345. 2, 149794, platform element, oral view. 3, 149795, lyriform element, oral view. 4, 149796, ambalodontan element, lateral view.

Figs 5–10. *Apsidognathus* sp. A. Shenxuanyi Member, Xuanhe Section, Guangyuan County, Sichuan, Sample Xuanhe 4. 5, 149797, platform element, oral view. 6, 149798, platform element, oral view. 7, 149799, astrognathodontan element, ?lateral view. 8, 149800, ambalodontan element, lateral view. 9–10, 149801, lyriform element, oral and lateral views.

Figs 11–14, 17–18. *Multicostatus dazhubaensis* Ding and Li, 1985. Shenxuanyi Member, Xuanhe Section, Guangyuan County, Sichuan, Sample Xuanhe 4. 11–12, 149802, lateral views. 13–14, 149803, posterior and anterior views. 17–18, 149804, lateral and oral views.

Figs 15–16. ?*Multicostatus dazhubaensis* Ding and Li, 1985. Shenxuanyi Member, Xuanhe Section, Guangyuan County, Sichuan, Sample Xuanhe 3. 149805, posterior and oral views.

Figs 19–20, ?21. *Multicostatus dazhubaensis* Ding and Li, 1985. Upper member, Xiushan Formation, Leijiatun Section, Shiqian County, Guizhou, Sample Shiqian 17. 19, 149806, posterior view. 20, 149807, oral view. 21, 149808, lateral view.

All figures ×60.

PLATE 13

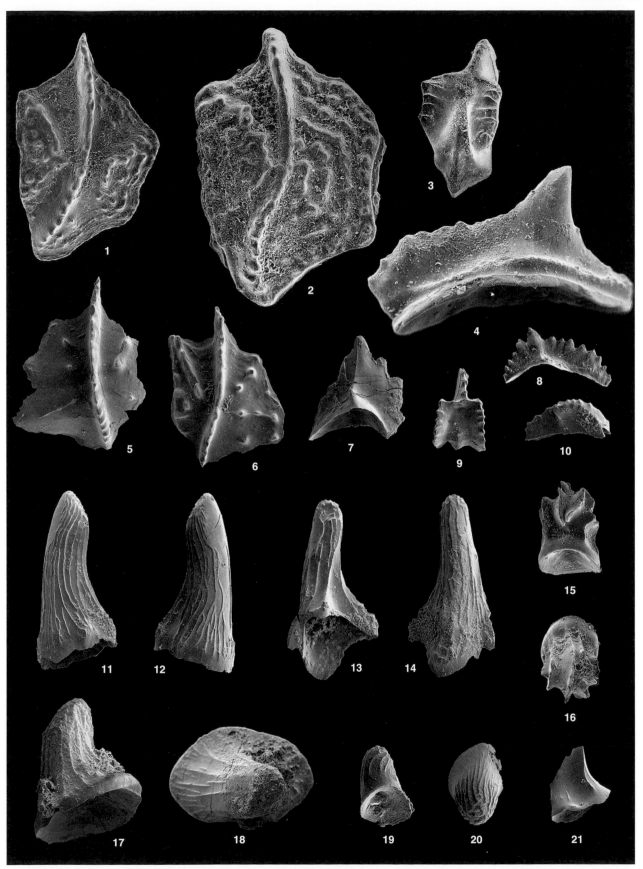

WANG and ALDRIDGE, *Apsidognathus, Multicostatus*

figs 10–11), and the rugose element is consistently more elongate (Pl.12, figs 12–14). It is possible that the former element does not belong to the *Apsidognathus* apparatus, but should be included in *Tuberocostadontus* (see below); the 'slender conical element' figured by Mabillard and Aldridge (1983, pl. 1, figs 12–13) also has similarities to specimens we suggest here to be part of the *Tuberocostadontus* apparatus. However, other characteristic *Tuberocostadontus* elements were not recognized in the collections from Wales reported by Mabillard and Aldridge (1983).

The platform elements from China and Wales also differ in several respects: the carina on the Chinese specimens is less strongly curved and the inner anterior ridge of nodes commonly more strongly developed; the lobes on the outer platform are generally less strongly developed (cf. Mabillard and Aldridge 1983, pl. 1, figs 1–2).

The suite of compressed elements illustrated and named by Ding and Li (1985) includes specimens close to those assigned here to *A. ruginosus scutatus*. There is no illustration of any platform element associated with these specimens, so it is impossible to be certain whether they belong to the same subspecies. If a platform element of *ruginosus* type were shown to be present at the same horizon as Ding and Li's specimens, then their names would have priority over *scutatus*, and a reviser could select a name appropriately. An even earlier name might be available in the compressed specimen referred to their new species *Parapygodus hemiorbicularis* by Zhou and Zhai (1983), but this specimen is broken and its complete shape is unknown; the nature of any associated platform element is also unknown.

Occurrence. Upper member, Xiushan Formation, Leijiatun section, Shiqian County, Guizhou; Yangpowan and Shenxuanyi members, Ningqiang Formation, Yushitan section, Ningqiang, Shaanxi; Shenxuanyi Member, Xuanhe section, Guangyuan, Sichuan; Daluzhai Formation, Huanggexi section, Daguan County, Yunnan.

Apsidognathus tuberculatus Walliser, 1964
Plate 13, figure 1

  *1964 *Apsidognathus tuberculatus* Walliser, p. 29, pl.5, fig.1; pl.12, figs 16–22; pl.13, figs 1–5 (platform element).
  1964 *Ambalodus galerus* Walliser, p. 27, pl. 6, fig.1; pl.12, figs 1–7 (ambalodontan element).
  1964 *Astrognathus tetractis* Walliser, p. 30, pl. 5, fig. 4; pl. 14, figs 1–2 (astrognathodontan element).
  1964 *Pygodus lyra* Walliser, p. 68, pl. 5, fig. 5; pl. 12, figs 8–14 (lyriform element).
  1964 ?*Pygodus lenticularis* Walliser, p. 67, pl. 4, fig. 17; pl. 12, fig. 15 (compressed element).

  1983 *Apsidognathus tuberoculatus* Walliser, 1964 [*sic*]; Zhou and Zhai, p. 268, pl. 65, fig. 5 (platform element).
 ?1987 *Ambalodus galerus* Walliser sf.; Ni, p. 390, pl. 62, fig. 2 (?ambalodontan element).
 ?1987 *Apsidognathus tuberculatus* Walliser sf.; Ni, p. 391, pl. 62, fig. 1 (platform element).
 ?1987 *Kailidontus typicus* Zhou, Zhai, Xian; Ni, p. 416, pl. 61, fig. 13 (astrognathodontan element).
 ?1987 *Pygodus? lyra* Walliser sf.; Ni, p. 435, pl. 62, fig. 11 (lyriform element).
 ?1988 *Astrognathus tetratis* Walliser [*sic*]; Wang G. X. *et al.*, pl. 1, fig. 7 (?astrognathodontan element)
  1988 *Apsidognathus tuberoculatus* Walliser [*sic*]; Wang G. X. *et al.*, pl. 1, fig. 11 (platform element).
  1988 *Kailidontus typicus* Zhou, Zhai et Xian; Wang G. X. *et al.*, pl. 1, fig. 16 (astrognathodontan element).
 ?1988 *Ambalodus galerus* Walliser; Wang G. X. *et al.*, pl. 1, fig. 21 (ambalodontan element).
v ?1990 *Apsidognathus tuberculatus tuberculatus* Walliser, 1964; Armstrong, p. 41, pl. 1, figs 12–16; pl. 2, figs 1–4 (multielement; with synonymy to 1987).
  1999 *Apsidognathus tuberculatus* Walliser, 1964; Melnikov, p. 69, pl. 20, figs 1–28 (multielement).

Diagnosis. See Armstrong (1990, p. 42), with the amendment that the lyriform element should always show raised lateral margins with transverse rugae.

Material. Platform, 8.

Remarks. Although Armstrong (1990) provided a full diagnosis and description, the illustrated specimens from Greenland differ sufficiently from those originally figured by Walliser (1964) to be referable to a separate subspecies, or even species; they are, therefore, included equivocally in synonymy here. The lyriform element from Greenland has a free blade and proximal lateral margins of the platform that are ornamented with a reticulate pattern, very distinct from any of Walliser's specimens. In addition, the platform element (Armstrong 1990, pl. 1, fig. 12) shows nodose ridges on the outer platform that do not all radiate from the centre of the element.

Melnikov (1999) illustrated a wide range of elements, including coniform morphologies, within his concept of the *A. tuberculatus* apparatus. The platform elements differ a little from Walliser's type suite and are in some respects more similar to the specimens from Greenland illustrated by Armstrong (1990), but a lyriform element is not illustrated for comparison.

Undoubted specimens of *A. tuberculatus* are uncommon in the material we have studied and tend to occur with other species of *Apsidognathus*, so the apparatus of the Chinese representatives is difficult to differentiate. It may be that, given the high morphological variability of the elements of *Apsidognathus*, these specimens represent extreme variants of other *Apsidognathus* species, and that species of this genus should only be identified on the basis of large populations.

Occurrence. Yangpowan Member, Ningqiang Formation, Yushitan section, Ningqiang, Shaanxi; Shenxuanyi Member, Xuanhe section, Guangyuan, Sichuan.

Apsidognathus aff. *arcticus* Armstrong, 1990
Plate 13, figures 2–4

aff. 1990 *Apsidognathus tuberculatus arcticus* Armstrong, p. 46, pl. 1, figs 1–11 (multielement).

Material. Platform, 1; ambalodontan, 1; lyriform, 1.

Remarks. Armstrong (1990) distinguished a subspecies *A. tuberculatus arcticus* on the basis of the characteristic lyriform element which is robust, strongly arched and bears prominent marginal ridges and a broad medial-posterior process. The platform element of the subspecies was considered to be indistinguishable from that of *A. tuberculatus tuberculatus*.

The lyriform element is very distinctive and markedly different from the specimens figured as *Pygodus lyra* by Walliser (1964, pl. 12, figs 8–14), which are regarded to be the lyriform elements of *A. tuberculatus*. We consider that these differences are strong enough to warrant specific status for *arcticus*.

Specimens from sample TT 345 from China include a lyriform element with the general characteristics of *A. arcticus*, but with ridges developed on the antero-lateral margins of the platform, rather than a reticular ornament. The associated platform element differs from typical specimens of *A. tuberculatus* in lacking radiating nodose ridges on the outer platform, although some specimens of *A. tuberculatus* also lack these (e.g. Walliser 1964, pl. 13, fig. 1). Given the limited number of specimens in the material we have studied, we cannot definitively separate them from *A. arcticus*, but the difference in the lyriform element might be diagnostic.

Occurrence. Ningqiang Formation, Dazhuba section, Ningqiang County, Shaanxi; Luomiang Formation, Kaili County, Guizhou.

Apsidognathus sp. A
Plate 13, figures 5–6, ?7–10

Material. Platform, 2; ambalodontan, 1 or more; lyriform, 1 or more; astrognathodontan, 1.

Remarks. Specimens of a platform element occur in sample Xuanhe 4 that show very limited ornamentation on the oral surface. Radiating rows of nodes occur on each platform lobe, but these may be reduced to one or two nodes in some rows, or even reduced to a low ridge; the surface between the nodes is completely smooth. The nodes appear to converge towards a single point on the gently curved carina. The platform outline most closely resembles that of *A. aulacis*, with the inner lobe narrower than the outer lobe and displaying a straight distal edge. These specimens may be extreme variants of a recognized species of *Apsidognathus*, but are sufficiently distinct to suggest that they represent a separate taxon. Other elements of the apparatus are difficult to assign, as specimens of *A. ruginosus* occur in the same sample. However, a distinctive astrognathiform element with two reduced processes is present (Pl. 13, fig. 7) that may belong to the same taxon, and lyriform elements in the sample are characteristically short, with a steep posterior face that lacks medial ridges (Pl. 13, figs 9–10). Associated ambalodontan elements (Pl. 13, fig. 8) vary from short to long, but are indistinguishable from those of other recognized species.

Occurrence. Shenxuanyi Member, Xuanhe section, Guangyuan County, Sichuan.

Genus PTEROSPATHODUS Walliser, 1964

1964 *Pterospathodus* Walliser, p. 66.
1964 *Carniodus* Walliser, p. 30.
1972 *Llandoverygnathus* Walliser, p. 72.
?1985 *Xainzadontus* Yu, p. 25.

Type species. *Pterospathodus amorphognathoides* Walliser, 1964, p. 66, by monotypy.

Diagnosis. Apparatus structure unknown, but includes at least three P elements, plus M, Sa, Sb and Sc. Pa element pastinate, sometimes appearing carminiscaphate; or pastiniscaphate or stelliscaphate with a restricted basal cavity. Pb element appears angulate with offset lobes to the basal cavity or anguliscaphate with narrow platform ledges and with downwardly projecting lips beneath cusp. Pc element pastinate. M element dolabrate with short adenticulate anterior and inner lateral processes. Sa element alate, all processes short and denticulate. Sb element tertiopedate, slightly asymmetrical to strongly asymmetrical, one lateral process may be adenticulate. Sc element dolabrate with anterior process directed downwards and backwards as an adenticulate anticusp. There may additionally be a further P element and a suite of elements with prominent cusps and short processes that often bear crowded denticles.

Remarks. The history of discussion of the composition of the *Pterospathodus* apparatus has been summarized by Männik (1998). A particular debate has revolved around the inclusion of varied elements originally referred by Walliser (1964) to species of the genus *Carniodus*. Männik (1998), following Jeppsson (1979), concluded from morphological and stratigraphical evidence that *Carniodus* elements were part of the *Pterospathodus* apparatus, and he suggested that the total apparatus had 14 element types. He recognized that this was more than is found in any apparatus known from a natural assemblage and suggested that some of the 'Carniodus' elements might have occurred as posterior process continuations of the S elements. This interpretation is supported by the pattern of growth of some *Pterospathodus* S elements noted by Walliser (1964) and further documented by Donoghue (1998); in these, there is clear fusion of units that initially grew independently with separate basal cavities (e.g. see Walliser 1964, fig. 4y, pl. 6, fig. 15; Donoghue 1998, fig. 7e–f). The positions of other elements are problematical, although it would not be expected that this apparatus has more than the nine S positions recognized throughout prioniodontid conodonts. However, recognition of the true nature of this complex apparatus probably awaits the discovery of a natural assemblage. In the meantime, the inclusion of some *Carniodus* elements must remain equivocal, although they most likely do belong. Here, we accept the assignment of elements of the type originally referred by Walliser (1964, p. 51, pl. 5, fig. 7, pl. 28, figs 12–18) to *Neoprioniodus subcarnus* to the Sc position of *Pterospathodus* (see Männik and Aldridge 1989, p. 895).

The homologies of the P elements with those of other balognathid apparatus are also uncertain. Three paired P elements, Pa, Pb and Pc, were recognized by Männik and Aldridge (1989), but Männik (1998) suggested that a fourth might be represented by specimens referred to ?*Carniodus carinthiacus* by Walliser (1964, p. 31, pl. 6, fig. 8, pl. 27, figs 20–26). A comparison with *Promissum pulchrum*, which has four pairs of P elements, P_{1-4} (Aldridge *et al.* 1995; Purnell *et al.* 2000), might suggest a similar pattern for *Pterospathodus*, but the P_1 and P_2 elements of *Promissum* are almost identical, which would not be the case in *Pterospathodus* if Männik's proposal were followed. It, therefore, remains possible that *Pterospathodus* had four morphologically distinct pairs of P elements, or just three morphologically distinct pairs (omitting ?*C. carinthiacus*), or four pairs of P elements with the P_1 and P_2 positions occupied by elements of similar morphology (i.e. with two Pa pairs and no ?*C. carinthiacus*).

The holotype of the type species of *Xainzadontus*, *X. dewukaxiaensis* Yu, 1985, is indistinguishable from ?*Carniodus carinthiacus* (see Walliser 1964, pl. 27, figs 20, 23), which is almost certainly an element from a *Pterospathodus* apparatus. *Xainzadontus* is, therefore, most probably a junior synonym of *Pterospathodus*.

The Pa elements of different species of *Pterospathodus* have been described as carminate or pastinate in form, but there have been uncertainties regarding the homologies between the processes on these elements and those of other early prioniodontids. Basal balognathids such as *Baltoniodus* have pastinate Pa elements with three primary processes conventionally regarded as anterior, posterior and lateral (Sweet 1981, fig. 10.1a–b). Thin section studies have shown that the element initially developed as a carminate juvenile with the posterior and 'lateral' processes forming the primary axis; the 'anterior' process developed at a slightly later ontogenetic stage (Viira *et al.* 2006, p. 226, fig. 5). Unpublished ontogenetic studies by Stephanie Curtis (née Barrett, 2000, PhD thesis, University of Leicester), using transmitted light and thin-sectioning (Text-fig. 13), indicate that the conventionally lateral process is suppressed in *Pterospathodus*, forming just a small expanded lobe on one side of the element. Thus, the long axis of *Pterospathodus* Pa elements is formed of the posterior and anterior processes of an essentially pastinate element; the carminate appearance of some species is a derived feature. The pennate and bifurcate processes developed by some taxa are secondary postero-lateral processes developed on the inner side of the element and are not homologous to the lateral process of *Baltoniodus*. Curtis also suggested that the Pb element of *Pterospathodus* is technically pastinate, with the third primary process represented by the outer lateral lobe beneath the cusp.

Pterospathodus amorphognathoides aff. *lennarti* Männik, 1998
Plate 14, figures 1–2

p 1992 *Spathognathodus pennatus procerus* Walliser, 1964; Qian *in* Jin *et al.*, p. 62, pl. 3, fig. 6 (*non* fig. 12a–b (?= *Pterospathodus procerus*)) (Pa element).
aff. 1998 *Pterospathodus amorphognathoides lennarti* Männik, p. 1019, pl. 3, figs 21–46, text-fig. 9 (multielement).

Material. Pa, 2.

Description. Pa element pastiniscaphate. Relatively large; blade straight or very slightly curved, with narrow ledge developed on better-preserved specimen. Outer later process short, lobate, with low axial ridge terminating distally as a small node. Inner lateral process bifurcate, axis offset posteriorly from outer lateral process, separated from blade by a low, arched, unornamented area; anterior branch relatively long with five axial nodes decreasing in

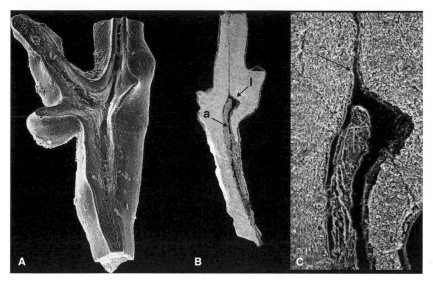

TEXT-FIG. 13. A. Aboral view of adult Pa element of *Pterospathodus amorphognathoides* (S. Curtis specimen 272, from Birches Farm Lane, Malvern Hills, Herefordshire, UK; see Aldridge 1972 for locality and horizon details), showing five branches to the basal cavity; ×35. B. Section through apex of basal cavity of adult Pa element of *P. amorphognathoides* (S. Curtis specimen 162, locality as above), in same orientation as A, showing that the primary processes are the 'anterior' (a) and 'lateral' (l) processes (arrowed); ×30. C. Close-up of apex of cavity in B, showing incipient development of 'posterior' process (arrowed); ×180. Photographs by Dr S. Curtis.

size distally, posterior branch short with a single, quite prominent node. Both branches with narrow ledges.

Other elements not recognized.

Remarks. These specimens have the characteristics of those referred to the *P. amorphognathoides* lineage by Männik (1998), but do not correspond directly to any of the subspecies he recognized. There is an overall similarity to the holotype of *P. a. lennarti* illustrated by Männik (1998, pl. 3, fig. 21), but that specimen has a sigmoidal blade, a lack of platform ledges on the inner lateral process and a low ridge connecting the inner lateral process to the blade. Männik (1998, p. 1019) considered the lack of a platform and the ridge between the blade and inner lateral process as diagnostic for *P. a. lennarti*. In *P. a. lithuanicus* Brazauskas, 1983, and in *P. a. amorphognathoides* Walliser, 1964, the proximal denticle of the bifurcate inner process is situated close to the main blade; the Pa element of *P. a. amorphognathoides* has strongly developed platform ledges on all processes.

Occurrence. Shenxuanyi Member, Xuanhe section, Guangyuan County, Sichuan (Xuanhe 6, TT 498).

Pterospathodus eopennatus Männik, 1998
Plate 14, figures 3–22

1981 *Spathognathodus celloni* Walliser; Zhou *et al.*, pl. 2, figs 12–13 (Pa element).

1981 *Ozarkodina adiutricis* Walliser; Zhou *et al.*, pl. 2, figs 22–23 (Pb element).

1983 *Pterospathodus pennatus* (Walliser, 1964); Zhou and Zhai, p. 295, pl. 68, fig. 1 (Pa element).

1983 *Spathognathodus celloni* Walliser, 1964; Zhou and Zhai, p. 295, pl.68, figs 3–4 (Pa element).

1983 *Ozarkodina adiutricis* Walliser, 1964; Zhou and Zhai, p. 287, pl. 67, figs 5–6 (Pb element).

1985 *Pterospathodus celloni* (Walliser, 1964); Qiu, p. 31, pl. 1, figs 1a–b, 2 (Pa element).

?1987 *Pterospathodus celloni* (Walliser); Ni, p. 434, pl. 61, figs 33, 36–37 (Pa element).

?1987 *Pterospathodus pennatus pennatus* (Walliser); Ni, p. 434, pl. 61, figs 31, 38 (Pa element).

1988 *Pterospathodus celloni* (Walliser); Qiu, pl. 1, fig. 3a-b (copy of Qiu 1985, pl. 1, fig. 1a–b) (Pa element).

1988 *Pterospathodus pennatus* (Walliser); Wang G.X. *et al.*, pl. 1, fig. 14 (Pa element).

?1988 *Spathognathodus celloni* Walliser; Wang G.X. *et al.*, pl. 1, fig. 17 (Pa element).

1992 *Spathognathodus celloni* Walliser; Qian *in* Jin *et al.*, p. 61, pl. 2, fig. 1a–b (Pa element).

1992 *Ozarkodina adiutricis* Walliser, 1964; Qian *in* Jin *et al.*, p. 60, pl. 2, fig. 3 (?Pa element).

1993 *Pterospathodus celloni* (Walliser); Xia, p. 210, pl. 3, fig. 13; pl. 4, figs 2–3 (Pa, Pb elements).

1993 *Pterospathodus pennatus pennatus* (Walliser); Xia, p. 211, pl. 4, fig. 129 (Pa element).

v 1996 *Pterospathodus celloni* (Walliser); Wang and Aldridge, pl. 5, figs 2–3 (Pa, Pb elements).

* 1998 *Pterospathodus eopennatus* Männik, p. 1007, pl. 1, figs 1–46, pl. 2, figs 23, 32–41, text-figs 4–6 (multielement; with synonymy to 1996).

v 2002 *Pterospathodus celloni* (Walliser); Aldridge and Wang, figs 66B–C. [copy of Wang and Aldridge 1996, pl. 5, figs 2–3] (Pa, Pb elements).

Diagnosis. See Männik (1998, p. 1009).

Material. Pa, 150; Pb, 138; Pc, 82; M, 76; Sa/b, 80; Sc, 32; carniodiform, 60; plus additional material from TT samples.

Remarks. As noted by Männik (1998), the Pa element is highly variable; denticles are relatively tall and development of a pennate lateral process is common. This contrasts with *P. celloni*, in which denticles are shorter on the posterior process than on the anterior, the lateral process is absent and narrow, but distinct ledges are usually developed on the posterior process (P. Männik, pers. comm. 2007). Specimens referred to *P. eopennatus* in the collections examined here are gently arched with 10–18 (mostly 10–15) denticles, highest in the middle of the blade; the cusp is sometimes prominent (e.g. Pl. 14, fig. 17). Pennate inner lateral processes are commonly developed on sinistral elements, more rarely on dextral elements (note that Männik and Aldridge (1989) transposed these elements as they considered the process to be outer lateral, rather than inner lateral; Männik (1998) correctly termed the process as inner lateral in the diagnosis, but then transposed sinistral and dextral in the description). Among the specimens illustrated by Männik (1998), these specimens compare most closely with those he referred to morph 1a, especially those in his figures 5R–S and 6B–C, but they are not identical, having lower anterior denticles and a more evident arching. In his description, Männik (1998, p. 1009) stated that elements of this morph are relatively long, with up to 16–18 denticles, but most of his figured specimens have fewer denticles than this. Männik (1998) differentiated two unnamed subspecies of *P. eopennatus*, based partly on the Pa morphs present; morph 1a occurs in both subspecies, as does morph 1b, which a few of our specimens resemble. There are also differences in the Pb ('Pb$_1$') elements, with that in subspecies 1 being diagnosed as relatively longer and more arched (Männik 1998, p. 1013); however, the long, arched specimen figured by Männik (1998, pl. 1, fig. 15) as a Pb$_1$ of subspecies 2 is very similar to some of the specimens we have studied (Pl. 14, fig. 7). At present, we do not feel able to assign the Chinese material to one or other of the subspecies. Morph 1 ranges throughout the *P. eopennatus* Biozone of Männik (1998, text-fig. 3).

Occurrence. Upper member, Xiushan Formation, Leijiatun section, Shiqian County, Guizhou; Yangpowan and Shenxuanyi members, Ningqiang Formation, Yushitan section, Ningqiang, Shaanxi; lower part of Shenxuanyi Member, Xuanhe section, Guangyuan, Sichuan.

Pterospathodus procerus (Walliser, 1964)
Plate 15, figures 1–3

*1964 *Spathognathodus pennatus procerus* Walliser, p. 80, pl. 15, figs 2–8 (Pa element).

EXPLANATION OF PLATE 14

Fig. 1. *Pterospathodus amorphognathoides* aff. *lennarti* Männik, 1998. Shenxuanyi Member, Xuanhe Section, Guangyuan County, Sichuan, Sample Xuanhe 6. 149809, Pa element, oral view.

Fig. 2. *Pterospathodus amorphognathoides* aff. *lennarti* Männik, 1998. Shenxuanyi Member, Xuanhe Section, Guangyuan County, Sichuan, Sample TT 498. 149810, Pa element, oral view.

Figs 3–16. *Pterospathodus eopennatus* Männik, 1998. Upper member, Xiushan Formation, Leijiatun Section, Shiqian County, Guizhou, Sample Shiqian 17. 3, 149811, Pa element, oral view. 4, 149812, Pa element, lateral view. 5, 149813, Pa element, lateral view. 6, 149814, Pa element, oral view. 7. 149815, Pb element, lateral view. 8, 149816, Sa/Sb element, posterior view. 9–10, 149817, Sa/Sb element, posterior view and close-up to show longitudinal striae on axial portion. 11, 149818, Pc element, lateral view. 12, 149819, M element, posterior view. 13, 149820, Sc element, outer lateral view. 14–16, 149821, 149822, 149823, carniodiform elements, lateral views.

Figs 17–20. *Pterospathodus eopennatus* Männik, 1998. Upper member, Xiushan Formation, Leijiatun Section, Shiqian County, Guizhou, Sample TT 740. 17, 20, 149824, Pa element, lateral and oral views. 18, 149825, Pa element, lateral view. 19, 149826, Pb element, lateral view.

Fig. 21. *Pterospathodus eopennatus* Männik, 1998. Upper member, Xiushan Formation, Leijiatun Section, Shiqian County, Guizhou, Sample TT 739. 149827, Pa element, upper view.

Fig. 22. *Pterospathodus eopennatus* Männik, 1998. Upper member, Xiushan Formation, Leijiatun Section, Shiqian County, Guizhou, Sample TT 720. 149828, Pa element, lateral view.

All figures ×80, except fig. 10, ×360.

PLATE 14

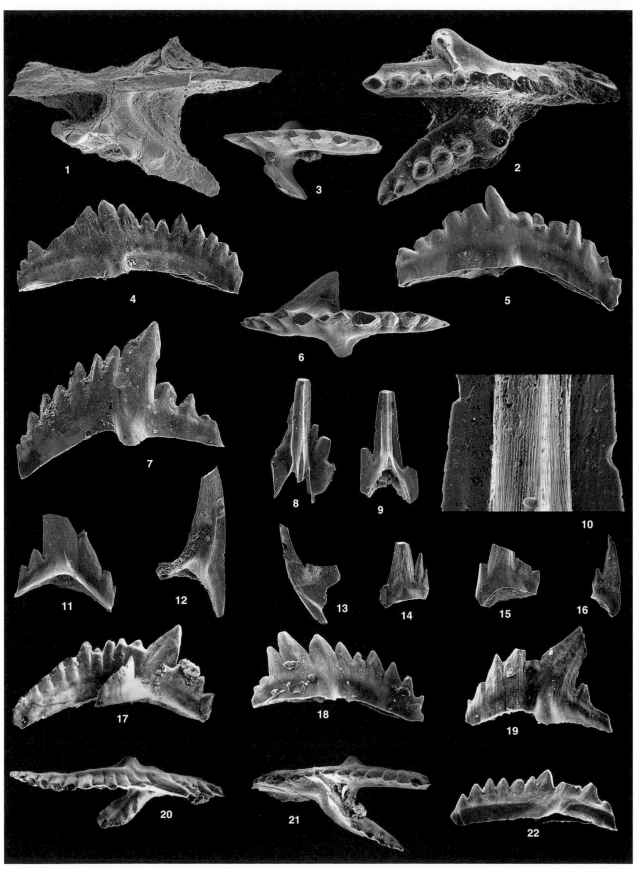

WANG and ALDRIDGE, *Pterospathodus*

1985 *Pterospathodus pennatus procerus* (Walliser); Yu, pl. 1, figs 1–2 (Pa element).

p 1987 *Pterospathodus pennatus procerus* (Walliser) 1964; An, p. 202, pl. 33, figs 5–6 only (Pa element).

p? 1993 *Pterospathodus celloni* (Walliser); Xia, pl. 4, figs 2a–c only (Pa element).

1997 *Pterospathodus procerus* (Walliser); Dumoulin *et al.*, fig. 4NN–PP (Pa, Pa, Pb elements).

1998 *Pterospathodus pennatus procerus* (Walliser, 1964); Männik, p. 1037, pl. 6, figs 1–25, 27–35, text-fig. 16 (multielement; with synonymy to 1998).

2001 *Pterospathodus pennatus procerus* (Walliser), 1964; Li and Qian, p. 97, pl. 1, figs 9–10 (Pa element).

2005 *Pterospathodus pennatus procerus* (Walliser, 1964); Jin *et al.*, pl. 1, fig. 4 (Pa element).

Material. Pa, 2; Pb, 3.

Remarks. The two specimens of the Pa element are closely similar to specimens figured by Walliser (1964, especially pl. 15, figs 2, 5); the blade is slightly arched and low centrally, and the blade and pennate inner lateral processes have narrow ledges. The Pa specimens illustrated by Männik (1998) are similar, but have stronger platform ledges. The Pb element (Pl. 15, fig. 2) associated with the Chinese Pa specimen is long with narrow platform ledges. The same sample (Xuanhe 1) also contains a large number of additional *Pterospathodus* elements that may or may not be from the same apparatus (Pl. 15, figs 4–17); they are all significantly smaller and include a shorter Pb element with a prominent cusp and without platform ledges (Pl. 15, fig. 4). A very small number of associated broken and very poorly preserved specimens may represent a smaller Pa element of a second *Pterospathodus* species. Given the uncertainties, all these additional elements are simply identified as *Pterospathodus* sp.

Given that a possible synonymy between *P. celloni* and *P. pennatus pennatus* was suggested by Männik and Aldridge (1989) and that Pa elements of *P. pennatus procerus* are clearly distinguished by the possession of platform ledges, retaining a subspecific status for *procerus* seems unfounded. Jeppsson (1997) and Dumoulin *et al.* (1997) considered it to be a distinct species, which we follow here.

Occurrence. Shenxuanyi Member, Xuanhe section, Guangyuan County, Sichuan (sample Xuanhe 1); Shenxuanyi Member, Ningqiang Formation, Ningqiang section, Ningqiang County, Shaanxi (sample TT 130).

Pterospathodus sinensis sp. nov.
Plate 15, figures 18–31

Derivation of name. Sino-, pertaining to China; -ensis, L., suffix denoting place.

Holotype. Specimen NIGPAS 149844 (Pl. 15, figs 18–19); Pa element.

Type locality and horizon. Xuanhe section, Guangyuan County, Sichuan; Shenxuanyi Member, sample Xuanhe 4.

Diagnosis. Pa element short, arched with prominent cusp; anterior denticles higher than posterior; lobes of basal cavity rounded in oral view, without marked offset on each side; lateral processes absent. Pb element arched, anterior process much higher than posterior, with denticle size on both processes decreasing gradually distally. Pc element with outer lateral costa becoming prominent and outwardly directed basally to form a triangular outline to the basal cavity in oral view.

Material. Pa, 25; Pb, 69; Pc, 27; M, 30; Sa/b, 7; Sc, 4; ?carniodiform, 3.

EXPLANATION OF PLATE 15

Figs 1–3. *Pterospathodus procerus* (Walliser, 1964). Shenxuanyi Member, Xuanhe Section, Guangyuan County, Sichuan, Sample Xuanhe 1. 1, 3, 149829, Pa element, lateral and oral views. 2, 149830, Pb element, lateral view.

Figs 4–17. *Pterospathodus* sp. Shenxuanyi Member, Xuanhe Section, Guangyuan County, Sichuan, Sample Xuanhe 1. 4, 149831, Pb element. lateral view. 5, 149832, Pc element, lateral view. 6, 149833, M element, posterior view. 7, 149834, Sc element, outer lateral view. 8, 149835, Sa/Sb element, posterior view. 9–10, 149836, carinthiaciform element, lateral and aboral views. 11–17, 149837, 149838, 149839, 149840, 149841, 149842, 149843, carniodiform elements, lateral views.

Figs. 18–31. *Pterospathodus sinensis* sp. nov. Shenxuanyi Member, Xuanhe Section, Guangyuan County, Sichuan, Sample Xuanhe 4. 18–19, 149844, Pa element, lateral and upper views (holotype). 20, 149845, Pa element, lateral view. 21, 149846, Pa element, lateral view. 22, 149847, Pb element, lateral view. 23, 149848, Pb element, lateral view. 24, 149849, Pc element, lateral view. 25, 149850, Pc element, lateral view. 26, 149851, M element, posterior view. 27, 31, 149852, Sc element, outer lateral view and close-up to show distribution of microstriae. 28, 149853, Sa/Sb element, lateral view. 29, 30, 149854, M element, posterior view and close-up to show striae.

Figures 1–3, ×70, all other figures ×80, except fig. 30, ×600, fig. 31, ×340.

PLATE 15

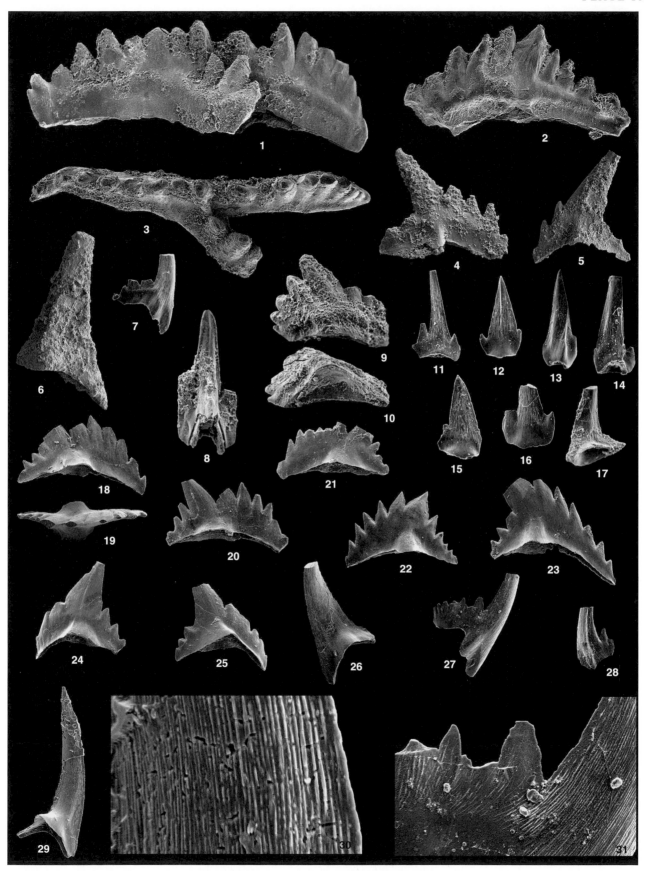

WANG and ALDRIDGE, *Pterospathodus*

Description. Pa element carminate, gently arched in lateral view, with short anterior and posterior processes and a prominent cusp. Cusp laterally compressed, posteriorly inclined, with sharp anterior and posterior edges. Anterior process with 4–7 erect, triangular denticles, fused basally; denticles of subequal size except for the distal one or two, which are usually smaller. Posterior process much lower and shorter, with 2–3 low denticles, generally discrete, although proximal denticle may be basally fused to cusp; denticles decrease in size distally. Basal cavity flaring most widely beneath anterior margin of cusp; lobes on both sides unornamented, inner lobe slightly wider and longer than outer lobe, lobes not offset or only slightly offset; cavity narrows gradually to anterior and posterior tips. White matter fills cusp and free tips of denticles. Cusp and denticles display subtle surface microstriae.

Pb element angulate, a little more strongly arched than Pa element, with a more-or-less prominent cusp. Cusp laterally compressed, posteriorly inclined, with sharp anterior and posterior edges. Anterior process high near cusp, decreasing in height distally and bearing 3–6 triangular denticles, which also decrease steadily in size distally. Posterior process much lower and shorter, with three or four low denticles, generally discrete, although proximal denticle may be basally fused to cusp; denticles decrease in size distally. Basal cavity flaring most widely beneath anterior margin of cusp; lobes on both sides unornamented, slightly offset and slightly pinched on inner side; cavity narrows gradually to anterior and posterior tips. White matter fills cusp and free tips of denticles. Cusp and denticles display subtle surface microstriae.

Pc element pastinate, with short denticulate anterior and posterior processes and an outer lateral costa. Cusp very prominent, tall and long, with sharp anterior and posterior edges. Basal edge of unit gently arched. Anterior process directed downwards and anteriorly, bearing 3–5 fused denticles. Posterior process slightly shorter or of similar length, directed less strongly downwards and bearing 3–5 more discrete, triangular denticles. Outer lateral costa arises from anterior margin of cusp, becoming more prominent and outwardly directed basally to form a triangular outline to the basal cavity in oral view. White matter fills cusp and free tips of denticles.

M element makellate, with a tall, strongly curved and slightly twisted cusp. Transverse section of cusp lenticular with sharp lateral edges. Convex lateral edge of cusp developed downwards into a short, adenticulate anticusp; concave edge gives rise to a short lateral process, adenticulate or bearing one or two incipient denticles. Posterior face of cusp expands posteriorly at base to give a prominent lip to basal cavity, which is shallow and extends as a broad groove to tips of processes. White matter deep, filling cusp above tip of basal cavity. Cusp with well-developed microstriae (Pl. 15, fig. 30), which are narrower and more densely crowded towards concave edge.

Sa/Sb element small, tertiopedate, with strong costae on cusp developed into short lateral and posterior processes bearing one or two small denticles.

Sc element dolabrate, with a tall, slender, gently curved cusp, a short anterior anticusp and a relatively short posterior process bearing about four laterally compressed, discrete denticles. Basal edge extended into a short aboral lip on outer side of cusp. White matter fills cusp and free tips of denticles. Cusp with well-developed microstriae; broader, more widely spaced microstriae present on the posterior process below the denticle row, but not extending to the denticle tips (Pl. 15, fig. 31).

Remarks. The Pa element differs from other species of *Pterospathodus* in lacking the characteristic offset of the lateral lobes of the basal cavity. However, the elements recognized closely resemble their counterparts in other species of the genus, so they are regarded as congeneric on the basis of the whole apparatus. The collection from sample Xuanhe 4 does not contain specimens of the kind referred by Männik (1998) to the Pb$_2$ position nor any carniodiform specimens, so their presence in the apparatus of *P. sinensis* remains undetermined.

Occurrence. Upper member, Xiushan Formation, Leijiatun section, Shiqian County, Guizhou (sample Shiqian 20); Shenxuanyi Member, Xuanhe section, Guangyuan County, Sichuan (sample Xuanhe 4).

?Family BALOGNATHIDAE Hass, 1959

Genus TUBEROCOSTADONTUS Zhou, Zhai and Xian, 1981

1981 *Costadontus* Zhou *et al.*, p. 131.
?1981 *Nericodina* Zhou *et al.*, p. 135.
1981 *Tuberocostadontus* Zhou *et al.*, p. 139.
1985 *Flatodus* Ding and Li, p. 16.
1986 *Pyrsognathus* Bischoff, p. 207.

Type species. *Tuberocostadontus shiqianensis* Zhou, Zhai and Xian, 1981, p. 139, by monotypy.

Original diagnosis. Conical form, surface of lower part ornamented with nodose ridges and of upper part with longitudinal striations (from Zhou *et al.* 1981, p. 131, translated by Zhen Yong-yi and B. G. Fordham).

Multielement diagnosis. Apparatus structure uncertain. ?Pa element modified segminiscaphate, subtriangular to rhombic in outline. ?Pb element anguliscaphate, with short processes. ?Pc element anguliscaphate with short processes and an inwardly expanded base bearing a row of low nodes. ?M element dolabrate, with longitudinal ridges variably developed on cusp and denticles. ?S elements with tall cusps and very short processes; cusps bear several strong costae. All elements with more or less distinct, smooth, crimp-like band near aboral margin, separated from distal portions by a ridge or platform ledge (emended from the diagnosis of *Pyrsognathus* Bischoff (1986, p. 207)).

Remarks. The composition of the apparatus of this genus is currently unclear. Fordham (1991, p. 29) considered that the apparatus of the type species included 'serratid', 'shiqianensid', 'Pb', 'Sa', 'Sb' and 'simplicid' elements, and that *Costadontus, Flatodus, Multicostatus, Nericodina* and *Pyrsognathus* were subjective junior synonyms. Elements referred to *Flatodus* are closely similar to those of *Tuberocostadontus*, and we agree that these two are synonymous. A suite of elements of the type assigned to *Costadontus* also occurs in association, and these specimens show similar longitudinal striations on the cusp. The assignment of *Nericodina* is more equivocal, as lenticular elements of this type may occur in the apparatus of either *Tuberocostadontus* or *Apsidognathus*. Elements comparable with *Multicostatus* are variable enough to suggest that they belong to a distinct apparatus (*contra* Wang and Aldridge 1998), and they are separated here, although the status and nature of *Multicostatus* remains problematic. The Chinese specimens assigned to *Tuberocostadontus* compare closely with the set of elements included by Bischoff (1986) in *Pyrsognathus*, although the species are different.

Additional elements in the Chinese collections that may be part of the *Tuberocostadontus* apparatus include a nearly symmetrical form (Pl. 16, figs 16–17). Fordham (1991, p. 30) included the 'slender conical element' of Mabillard and Aldridge (1983, pl. 1, fig. 12) in synonymy with *T. shiqianensis*, and we also include broadly similar elements in the apparatus here (Pl. 16, figs 5–6, 18–19, 22–23), although the specimens from Wales probably come from a distinct species.

The familial assignment of *Tuberocostadontus* is highly equivocal, but we provisionally include the genus in the Balognathidae on the basis of the presence of compressed elements similar to those of *Apsidognathus* and on the possible presence of three pairs of P elements.

Tuberocostadontus spp.
Plate 16, figs 1–35

Material. ?Pa, 53; ?Pb, 31; ?Pc, 7; ?M, 12; ?Sa, 1; ?Sb-Sd, 203; plus additional material from TT samples.

Remarks. The specimens of *T. shiqianensis* illustrated by Zhou *et al.* (1981, pl. 1, figs 48–52), including the holotype, show well-developed tuberculation around the basal margin that is not evident in the specimens in the collections we have studied, so *shiqianensis* is probably not an appropriate name for the taxa we illustrate. The suites of specimens in our illustrations come from two principal samples: Xuanhe 3 (Pl. 16, figs 1–17) and Shiqian 20 (Pl. 16, figs 20–35). Some elements in these two collections are closely comparable, but others differ, suggesting that two species might be represented. Available names for these species include those derived from the two species of *Costadontus* named and described by Zhou *et al.* (1981, p. 131): *C. saggitodontoides* (type species) and *C. serratus.* Specimens closely similar to those of *C. serratus* (see Zhou *et al.* 1981, pl. 1, figs 3–5) occur in both Xuanhe 3 (Pl. 16, figs 8–11) and Shiqian 20 (Pl. 16, figs 31–32). None of our specimens directly match *C. sagittodontoides* (see Zhou *et al.* 1981, p. 131, fig. 3, pl. 1, figs 6–9), although two in Shiqian 20 have similarities (Pl. 16, figs 24–26). In view of the uncertainties in apparatus structure and in appropriate name, we provisionally describe two species of *Tuberocostadontus* in open nomenclature and do not attempt to construct synonymy lists. Resolution of these problems may rely on an examination of type specimens and their associated elements.

Tuberocostadontus sp. A
Plate 16, figures 1–15, ?16–17

Description. ?Pa element very compressed, with inner lateral face less strongly developed than outer lateral face. Denticle row strongly arched, forming an acute angle, with a distinct cusp, directed very slightly posteriorly. Anterior process steep, adenticulate; posterior process with fused low denticles forming a serrate ridge. Inner lateral face of element mostly smooth, with two or three medial striae and a broad, smooth basal margin; basal edge gently sinuous. Outer lateral face with impersistent low ridges semi-parallel to basal edge, sometimes bifurcating to produce a hint of a reticular pattern.

?Pb element anguliscaphate. Cusp prominent, inclined posteriorly, with a lenticular transverse section and smooth lateral faces. Anterior process short, directed strongly downwards, with two incipient denticles; posterior process also short, directed less strongly downwards with two incipient denticles. Basal cavity lip flares strongly on inner side and narrowly on outer side, unornamented.

?Pc element scaphate, with prominent cusp and broad base. Cusp erect, laterally compressed with sharp anterior and posterior edges. Two low primary processes, anterior and posterior but indistinguishable, arise from the edges of the cusp and bear very low denticles that form a serrated ridge. Secondary lateral process developed on inner side as two low nodes. Unit with a smooth basal margin, nearly vertical, similar to that of ?Pa element. Basal cavity flared gently outwards and more strongly inwards to give a subtriangular outline in oral view.

?M element makellate. Cusp erect, somewhat twisted, subtriangular in transverse section, with a gently convex anterior face and sharp lateral and posterior edges. Inner lateral process adenticulate with a sharp oral edge; short, directed laterally and slightly downwards. Outer lateral process with about three low, triangular denticles, decreasing in height distally. One or two faint longitudinal ridges present on the posterior face of each process, near junction with cusp. A smooth basal margin is separated from the anterior and posterior faces by a slight ridge. Basal cavity broad beneath cusp, narrowing gradually beneath processes.

?Sa element alate. Single symmetrical specimen, with small cusp and deep lateral processes, the distal ends of which are directed downwards to form a gentle basal arch. Both processes adenticulate. Anterior and posterior faces with fairly prominent medial ridge, adjacent to which are faint, short longitudinal ridges on the process faces. Narrow, smooth margin developed basally on both processes. Basal cavity narrow, widest centrally and narrowing beneath processes.

?Sb–d elements comprise a suite of elements, including forms referred to *Costadontus* by previous authors (see also Bischoff 1986, for *Pyrsognathus*). These all have very prominent cusps and short processes, which may be adenticulate or bear low denticles fused to form undulose ridges. Cusps and processes characterized by longitudinal ridges of different lengths; none of the ridges extends to the basal edge and there is a smooth basal margin on all specimens. Some specimens have anterior and posterior processes (Pl. 16, fig. 8), some have only an anterior process (Pl. 16, figs 9–11), some have an anterior ridge and a short, poorly denticulate posterior process (Pl. 16, figs 12–13), and some have a short posterior process and one short lateral process (Pl. 16, figs 14–15). All elements with a broad, deep basal cavity.

Tuberocostadontus sp. B
Plate 16, figures 18–35

Description. ?Pa element very compressed, with inner lateral face less strongly developed than outer lateral face. Cusp prominent to very prominent, erect, with two or three longitudinal striae on each face. Anterior process steep and short, adenticulate but with a slightly undulose oral edge; posterior process a little longer, also undulose but adenticulate. Inner lateral face of element with a broad, smooth basal margin; basal edge concave. Outer lateral face with two longitudinal striae becoming more prominent

towards the base, the more medial one bifurcating near the base but not reaching basal edge. These specimens are similar to that illustrated as *Flatodus unicostata* Ding and Li (1985, p. 16, pl. 1, figs 20–21), type species of their genus *Flatodus*.

?Pb element scaphate. Cusp not prominent, inclined posteriorly, with a lenticular transverse section at the tip. Anterior process short and deep, with two or more denticles, which are nearly as high as the cusp proximally. Posterior face of cusp with sharp ridge that bifurcates downwards into short, low posterior and postero–lateral processes, which are adenticulate or bear a single low denticle. Outer lateral face of element with several longitudinal ridges of different lengths, none reaching basal edge. Smooth basal margin separated from faces of element by a well-defined ridge parallel to basal edge. Basal cavity deep, with lip flaring strongly on inner side and more narrowly on outer side.

?Pc element not recognized.

?M element makellate. Cusp erect, a little twisted, ovoid in transverse section, with sharp lateral edges and a medial ridge posteriorly. Inner lateral process adenticulate with a sharp oral edge; short, directed laterally and slightly downwards. Outer lateral process short, directed more strongly downwards with an undulose oral edge. Longitudinal ridges parallel the medial ridge laterally, of different lengths, but none reaching the base. Smooth basal margin developed on anterior and posterior faces. Basal cavity relatively broad beneath cusp, narrowing gradually beneath processes.

?Sa element of the type described for *T.* sp. A not recognized.

?Sb–d elements. As for *T.* sp. A there is a suite of elements that includes some forms referred to *Costadontus* by previous authors. These all have very prominent cusps and short processes, which are adenticulate but have undulose oral edges. Cusps and processes characterized by longitudinal ridges of different lengths; none of the ridges extends to the basal edge, and there is a smooth basal margin on all specimens (e.g. Pl. 16, fig. 35). Some specimens have very short anterior and posterior processes (Pl. 16, figs 29–30); these resemble those assigned to *Flatodus simplex* Ding and Li (1985, p. 16, pl. 1, figs 18–19), but have taller cusps. Other specimens have only an anterior process,

---

EXPLANATION OF PLATE 16

Figs 1–15. *Tuberocostadontus* sp. A. Shenxuanyi Member, Xuanhe Section, Guangyuan County, Sichuan, Sample Xuanhe 3. 1–2, 149855, ?Pa element, inner lateral and outer lateral views. 3–4, 149856, ?Pc element lateral and oral views. 5–6, 149857, ?Pb element, lateral and oral views. 7, 149858, ?M element, posterior view. 8–9, 149859, ?S element, lateral views. 10–11, 149860, ?S element, lateral views. 12–13, 149861, ?S element, lateral views. 14–15, 149862, ?S element, lateral views.

Figs 16–17. *Tuberocostadontus* sp. A? Shenxuanyi Member, Xuanhe Section, Guangyuan County, Sichuan, Sample Xuanhe 3. 149863, ?Sa element, posterior and anterior views.

Figs 18–19. *Tuberocostadontus* sp. B. Shenxuanyi Member, Xuanhe Section, Guangyuan County, Sichuan, Sample Xuanhe 2. 149864, ?Pb element, lateral and oral views.

Figs 20–35. *Tuberocostadontus* sp. B. Upper member, Xiushan Formation, Leijiatun Section, Shiqian County, Guizhou, Sample Shiqian 20. 20–21, 149865, ?Pa element, inner lateral and outer lateral views. 22–23, 149866, ?Pb element, posterior and oral views. 24–25, 149867, ?S element, lateral views. 26, 149868, ?M element, posterior view. 27, 149869, ?S element, lateral view. 28, 149870, ?S element, lateral view. 29–30, 149871, ?S element, lateral views. 31–32, 149872, ?S element, lateral views. 33–35, 149873, ?S element, lateral views and close-up to show smooth basal margin.

All figures ×60, except fig. 35, ×300.

PLATE 16

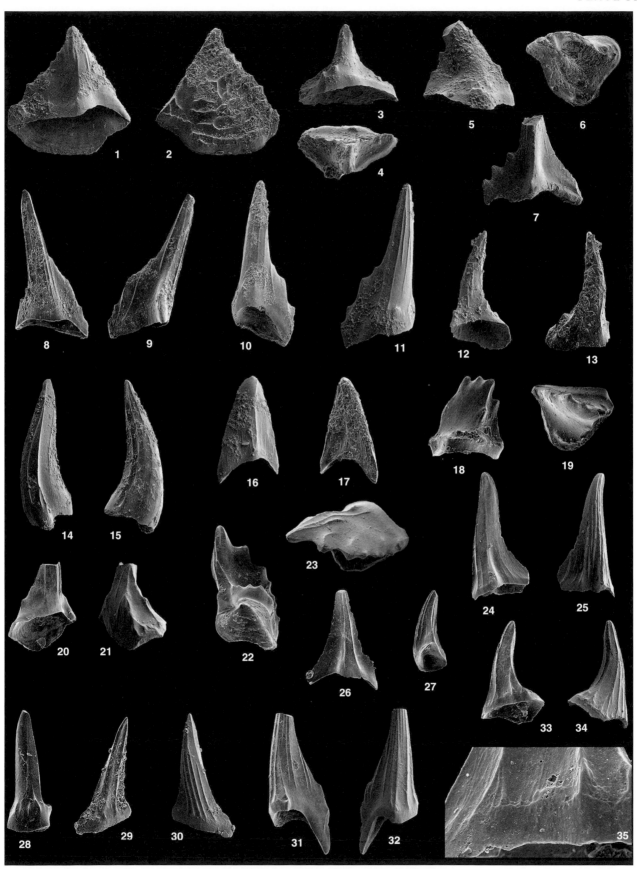

WANG and ALDRIDGE, *Tuberocostadontus*

which may be directed strongly downwards (Pl. 16, figs 31–32), some have an anterior ridge and a short, poorly denticulate posterior process (Pl. 16, figs 24–25, 33–34), and some have a short posterior process and short lateral processes (Pl. 16, fig. 27). One specimen (Pl. 16, fig. 28) has an erect, compressed cusp with sharp edges and medial ridges on each face and a narrowly expanded base with small tubercles above the smooth basal margin. All elements, except that in Pl. 16, figs 29–30, with a broad, deep basal cavity. In general, the S elements of *T.* sp. B have more strongly developed longitudinal ridges than those of *T.* sp. A, but not as prominent as in the specimens illustrated by Bischoff (1986, pls 32–34) or in the specimens of *T. shiqianensis* illustrated by Zhou *et al.* (1981, pl. 1, figs 48–52).

Occurrence. Upper member, Xiushan Formation, Leijiatun section, Shiqian County, Guizhou; Yangpowan and Shenxuanyi members, Ningqiang Formation, Yushitan section, Ningqiang, Shaanxi; Shenxuanyi Member, Xuanhe section, Guangyuan, Sichuan.

Family GAMACHIGNATHIDAE fam. nov.

Remarks. Sweet (1988) included the Upper Ordovician genus *Gamachignathus* McCracken *et al.*, 1980, in the family Balognathidae, but there are distinct morphological differences in all elements from other genera referred to that family. The placement within the same clade as other balognathids is not supported by the cladistic analysis of Donoghue *et al.* (2008), and the distinctive Pc?, M and S elements set *Gamachignathus* apart. There is, however, a similarity in element morphology and in apparatus structure within the genera *Gamachignathus*, *Birksfeldia* Orchard 1980 and *Galerodus*, and we suggest here that they should be assigned to a single, separate family.

The cladistic analysis by Donoghue *et al.* (2008) placed *Gamachignathus* as a basal member of the Order Ozarkodinida. However, McCracken *et al.* (1980) originally reconstructed *Gamachignathus* as having eight paired elements within the apparatus, of which two were assigned to e (= M) locations; Aldridge *et al.* (1995, p. 288) suggested that it was more likely that one of these is a Pc or Pd (P_3 or P_4) element. The characters for presence of P_3 and P_4 elements were scored as '?' for *Gamachignathus* by Donoghue *et al.* (2008, table 1); pending resolution of the number of P elements, the family Gamichignathidae is provisionally placed here outside the Ozarkodinida, as that order is diagnosed as containing apparatuses with only two P positions (see below).

Genus GALERODUS Melnikov *in* Tsyganko and Chermnyh, 1987

Type species. Galerodus magalius Melnikov *in* Tsyganko and Chermnyh, 1987, by monotypy.

Remarks. The genus *Galerodus* was erected by Melnikov (*in* Tsyganko and Chermnyh 1987, pl. 26, fig. 8) on the basis of a P element that was illustrated, but not described. Melnikov (1999, p. 71, pl. 21, figs 13–32) refigured the same specimen (fig. 31) together with a number of other P elements showing a wide morphological variation, but sharing a single row of denticles, gently sigmoidal in oral view, and an open basal cavity along much of the length of the unit. Melnikov (1999) did not illustrate other elements that might be assigned to the same apparatus. Zhou *et al.* (1981, pl. 2, figs 28–29) figured a similar type of P element as a new species, *Aphelognathus macroexcavatus*; a number of other specimens figured by them as separate form taxa are now considered to be part of the same apparatus (see below). *Aphelognathus* does not appear to be an appropriate generic assignment for this apparatus, as the apparatus of *Aphelognathus*, as exemplified by *A. kimms-wickensis* Sweet, Thompson and Satterfield (1975, pp. 31–33, pl. 2, figs 18–23) contains ramiform elements with robust denticles and flared basal cavities; Wang and Aldridge (1996, pl. 4, figs 1–6) provisionally assigned the Chinese material to *Gamachignathus? macroexcavatus* (Zhou *et al.*). Bischoff (1986, p. 190, pl. 28, figs 13–33) illustrated a new species, *Pterospathodus cadiaensis*, with very similar P elements to those from China; other elements he assigned to the apparatus are also comparable to some of the other elements of the Chinese apparatus. However, the type species of *Pterospathodus*, *P. amorphognathoides* Walliser, 1964, has a different apparatus, which includes pastinate Pb and Pc elements and a triangulate M element (see Männik and Aldridge 1989, text-fig. 1G–L), probably together with a range of 'carniodiform' S elements (see Männik 1998 and above). It seems that *Galerodus* provides the most appropriate generic assignment for the taxon *A. macroexcavatus*.

Galerodus macroexcavatus (Zhou, Zhai and Xian, 1981)
Plate 17, figs 1–19

1981 *Aphelognathus macroexcavatus* Zhou *et al.*, p. 130, pl. 2, figs 28–29 (Pa element).

?1981 *Exochognathus orbicudentatus* Zhou *et al.*, p. 132, pl. 1, fig. 15 (Sb element).

1981 *Exochognathus luomianensis* Zhou *et al.*, p. 132, pl. 1, figs 16–17 (Sb element).

1981 *Hibbardella luomianensis* Zhou *et al.*, p. 133, pl. 1, figs 18–19 (Sa element).

?1981 *Neoprioniodus longibaris* Zhou *et al.*, p. 134, pl. 1, figs 30–31 (M element).

1981 *Paracordylodus guizhouensis* Zhou *et al.*, p. 136, pl. 1, fig. 44 (Sc element).

1983 *Aphelognathus macrodentatus* Zhou, Zhai et Xian, 1981; Zhou and Zhai, p. 268, pl. 65, fig. 4a–b (Pa element).

1983 *Paracordylodus guizhouensis* Zhou, Zhai et Xian, 1981; Zhou and Zhai, p. 291, pl. 68, fig. 25 (Sc element).

1983 *Exochognathus luomianensis* Zhou, Zhai et Xian, 1981; Zhou and Zhai, p. 275, pl. 65, figs 25–26 (Sb element).

?1983 *Exochognathus orbicudentatus* Zhou, Zhai et Xian, 1981; Zhou and Zhai, p. 276, pl. 65, fig. 28 (Sb element).

?1983 *Neoprioniodus longibaris* Zhou, Zhai et Xian; Zhou and Zhai, p. 283, pl. 66, fig. 22 (M element).

1986 *Pterospathodus cadiaensis* Bischoff, p. 190, pl. 28, figs 13–33 (multielement).

1987 *Exochognathus* sp. An, pl. 31, figs 7–10 (Sb element).

?1989 *Aphelognathus* sp. A Yu *in* Jin *et al.*, pl. 6, figs 10–11; pl. 7, fig.4 (Pa element).

1989 *Paracordylodus guizhouensis* Zhou, Zhai et Xian, 1981; Yu *in* Jin *et al.*, p. 106, pl. 5, fig. 7; pl. 6, fig. 13 (Sc element).

1989 *Exochognathus luomianensis* Zhou, Zhai et Xian; Yu *in* Jin *et al.*, pl. 5, figs 4, 6 (Sb element).

p 1989 *Ligonodina egregia* Walliser, 1964; Yu *in* Jin *et al.*, p. 103, pl. 7, fig. 13 only (Sc element).

v 1996 *Gamachignathus? macroexcavatus* (Zhou *et al.*); Wang and Aldridge, pl. 4, figs 1–6 (multielement).

v 2002 *Gamachignathus? macroexcavatus* (Zhou *et al.*); Aldridge and Wang, fig. 65A–F [copy of Wang and Aldridge 1996, pl. 4, figs 1–6] (multielement).

Emended diagnosis. Apparatus with six distinct elements. Pa element very variable, gently arched or sigmoidal in lateral view and gently bowed or sigmoidal in oral view, with a wide, open basal cavity that may flare laterally beneath an indistinct cusp; oral edge bears 6–12 erect, triangular denticles with white matter blocks filling the free tips. Pc? element with a triangular cusp and with short posterior and antero-lateral processes on large specimens. M element dolabrate with a broad, high cusp and an adenticulate or weakly denticulate anticusp; two morphological groups can be distinguished and may represent different locations. Sa element alate with downwardly directed lateral processes and a long, curved posterior process with denticles that increase in size distally. Sb element tertiopedate, with one long, adenticulate, lateral process, a longer denticulate lateral process and a short posterior process. Sc element very characteristic, with a very tall cusp, a very long, adenticulate anticusp and a denticulate posterior process; basal profile between the two processes markedly U-shaped.

Material. Pa, 425; Pc?, 157; M, 534; Sa, 71; Sb, 323; Sc, 426; plus additional material from TT samples.

Description. Pa (and Pb?) element carminiscaphate. Blade nearly straight, gently curved or slightly sigmoidal in oral view, and gently arched to sigmoidal in lateral view. Blade bears 6–12 tri-angular, laterally compressed denticles with cusp inconspicuous. Denticulation very variable; denticles may be of subequal size, posterior denticles may be larger than the anterior ones, or single denticles at any point may be larger than their neighbours. Basal cavity wide throughout entire length of element, maybe flaring gently beneath cusp and, in some specimens, at anterior end and/or at posterior end. White matter fills free tips of denticles.

Pc? element bipennate. Cusp tall and strongly compressed laterally. Posterior process very short and may bear a tiny, rudimentary, laterally compressed denticle. In juvenile specimens, anterior process barely developed, but adult forms with a short antero-lateral process, which is directed sharply downwards and inwards and bears up to five small, discrete denticles. Basal cavity flares beneath cusp but narrows abruptly to a slit beneath processes. Entire cusp and tips of denticles, if present, filled with white matter.

M element dolabrate with a moderate to tall cusp, laterally compressed and curved posteriorly and inwards. Anterior and posterior edges of cusp sharp, outer lateral face gently convex, inner lateral face convex axially and becoming flattened towards the margins. Anterior margin extended downwards to form an anticusp, which may be directed straight downwards, or downwards and anteriorly. Anticusp commonly adenticulate, but may bear up to five small denticles; distal termination of process very sharp. Posterior process may be short, occasionally adenticulate but more commonly with at least one laterally compressed denticle, or it may be long and bear up to eight even-sized denticles. Base deeply excavated beneath cusp, with a strongly inwardly flaring lip, and continues as a narrow groove beneath processes. Microstriae well developed on cusp. The variation in orientation of the anticusp and in the development of denticulation may indicate that more than one element location is represented by these specimens, but the variation is continuous, and it is not possible to separate two morphological types.

Sa element alate, with a slender, posteriorly curved cusp bearing two prominent antero-lateral costae that extend directly downwards into two lateral processes that form a narrow, symmetrical arch. Each lateral process bears up to three discrete, antero-posteriorly compressed denticles. Posterior margin of cusp sharp and extending into a long posterior process, which is laterally compressed, tall, and smoothly curved posteriorly and downwards. Posterior process with up to five discrete, laterally compressed denticles, which increase in size distally; the distal two denticles may be conspicuously large and flat. Basal cavity narrow, slit-like beneath each process, not flaring markedly beneath cusp. Cusp and all denticle tips composed of white matter.

Sb element tertiopedate, with a tall cusp that is reclined posteriorly and somewhat twisted; transverse section of cusp lenticular near tip, but a posterior costa is developed proximally. Sharp lateral edges of cusp give rise to two asymmetrically disposed lateral processes; one nearly as long as cusp, adenticulate and directed obliquely downwards, its adenticulate edge forming a sharp ridge which continues in a slightly concave line from the lateral costa. Other lateral process curved downwards and bearing up to six antero-posteriorly compressed, discrete denticles that generally increase in size distally.

Posterior process, arising from posterior costa of cusp, short and directed downwards, adenticulate or bearing one or two discrete, laterally compressed denticles. Basal cavity very narrow beneath the processes and not flared at all beneath cusp. Cusp, denticle tips and upper portions of ridges on the processes composed of white matter.

Sc element dolabrate. Cusp very long, plano-convex in transverse section with inner face flattened, and posteriorly curved. Anterior edge of cusp sharp and slightly bowed inwards, confluent with a very long adenticulate anticusp, which narrows very steadily distally; distal extremity of anticusp broken away in all specimens. Posterior process curves gently downwards and bears three to nine laterally compressed denticles of more or less even size. Lower edges of the two processes produce a profile that is asymmetrically U-shaped. Basal cavity is a narrow slit, extending along both processes and showing no flaring beneath cusp. Cusp, denticle tips and anterior margin of anticusp are composed of white matter. Microstriae well developed on cusp, but absent from denticles (Pl. 17, fig. 19).

Remarks. Specimens of the Pa element are variable in denticulation and degree of arching, and it may be that both the Pa and Pb positions are represented by specimens referred to the Pa position here. Alternatively, there may only be two P positions in this apparatus. The element we tentatively designate to the Pc position is delicate and easily broken; hence, it is probably underrepresented in recovered collections.

Bischoff (1986) largely reconstructed the apparatus of a similar species, which he referred to a new taxon, *Pterospathodus cadiaensis*; we consider this to be another species of *Galerodus*. The Pa elements he illustrated (Bischoff 1986, pl. 28, figs 13, 17–22, 25) are variable, but show a very similar range of morphology to those from China described above, although the denticles are generally less widely spaced. One of the specimens he designated as a Pb element (Bischoff 1986, pl. 28, fig. 26) is very broken and might also be a Pa element. His other illustrated Pb elements (Bischoff 1986, pl. 28, figs 31, 32) also appear to be a little broken, but show some resemblance to the specimens assigned to the Pc? position here, differing in

the larger size of the denticle on the antero-lateral process and in the presence of denticulation on the posterior process. The specimens he referred to the M position (Bischoff 1986, pl. 28, figs 15, 27–29) we regard as being Sc elements and those he designated as S elements (Bischoff 1986, pl. 28, figs 16, 30, 33) are here distinguished as Sb elements. The Sc elements of *G. cadiaensis* have a greater angle between the anticusp and posterior process than those of *G. macroexcavatus*, resulting in a less U-shaped profile to their lower edges. Bischoff did not recognize elements comparable to those we regard as the M and Sa elements of *G. macroexcavatus*.

The Pa elements assigned by Melnikov (*in* Tsyganko and Chermnyh 1987; Melnikov 1999) to *Galerodus magalius* differ from those of *G. macroexcavatus* in the closer packing of the denticles, a feature they share with *G. cadiaensis*. It is possible that *magalius* and *cadiaensis* are synonyms.

There are similarities between the apparatus of *G. macroexcavatus* and that of the Late Ordovician *Gamachignathus ensifer* (see McCracken *et al.* 1980, pl. 10.1, figs 1–17), type species of the genus *Gamachignathus* McCracken *et al.* 1980. These are particularly evident in the elements assigned by McCracken *et al.* (1980) to the e-2 (Pc?), e-1 (M), b (Sb) and a-1 (Sc) positions. However, there are differences in the morphology of the elements they assigned to the equivalent of P positions, especially the f (= Pb) element, which is pastinate with three denticulate processes and has no counterpart in *G. macroexcavatus*. The c (= Sa) element is also very different, lacking the long, downcurved posterior process. Wang and Aldridge (1996) equivocally referred *macroexcavatus* to *Gamachignathus* to highlight the similarities, but we now consider that the species should be assigned to a separate genus.

Occurrence. Leijiatun Formation and lower and upper members, Xiushan Formation, Leijiatun section, Shiqian County, Guizhou; Wangjiawan Formation and Yangpowan and Shenxuanyi

---

Figs 1–19. *Galerodus macroexcavatus* (Zhou *et al.*, 1981). Upper member, Xiushan Formation, Leijiatun Section, Shiqian County, Guizhou, Sample Shiqian 17. 1–2, 149874, Pa element, lateral and oral views. 3, 149875, Pa element, lateral view. 4, 149876, Pa element, lateral view. 5, 149877, Pa element, lateral view. 6, 149878, M element, lateral view. 7, 149879, Pc? element, oblique lateral view. 8, 149880, Pc? element, lateral view. 9, 149881, M element lateral view. 10, 149882, M element, lateral view. 11, 149883, Sa element, lateral view. 12, 149884, Sa element, oblique lateral view. 13, 149885, Sa element, lateral view. 14, 149886, Sc element, inner lateral view. 15, 149887, Sb element, posterior view. 16, 149888, Sb element, posterior view. 17, 149889, Sc element, inner lateral view. 18–19, 149890, Sc element, outer lateral view and close-up to show distribution of striae.

Fig. 20. *Galerodus*? sp. Shenxuanyi Member, Xuanhe Section, Guangyuan County, Sichuan, Sample Xuanhe 5. 149891, Pa element, lateral view.

All figures ×100, except fig. 19, ×550.

PLATE 17

WANG and ALDRIDGE, *Galerodus*

members, Ningqiang Formation, Yushitan section, Ningqiang, Shaanxi; Shenxuanyi Member, Xuanhe section, Guangyuan, Sichuan.

Galerodus? sp.
Plate 17, figure 20

Remarks. A single specimen of a Pa element from the Shenxuanyi Member (sample Xuanhe 5) resembles the Pa of *G. macroexcavatus*, but has denticles that are more slender and peg-like, especially at the posterior end; the spaces between the denticles are also U-shaped and are broad in the central part of the specimen.

Order OZARKODINIDA Dzik, 1976

Diagnosis. Apparatus with 15 elements: two pairs of P elements, one pair of M elements and nine S elements. P_1 elements carminate, stellate, segminate, bipennate or with inner and outer lateral processes. P_2 element angulate, bipennate or with inner and outer lateral processes. S_1 element with inner lateral and outer lateral processes. Processes on all elements denticulate.

Remarks. The cladistic study by Donoghue *et al.* (2008) united taxa previously separated as ozarkodinids and prioniodinids in a single order.

Suborder PRIONIODININA Sweet, 1988

Diagnosis. Ozarkodinid conodonts with an inner lateral process on the Pb element. Denticles characteristically peg-like.

Remarks. Prioniodinins differ from members of the Suborder Ozarkodinina primarily in the disposition of the primary processes on the P elements; using the standard landmark of the orientation of the cusp, the processes on prioniodinins are lateral and those of ozarkodinins are anterior and posterior (see Donoghue *et al.* 2008). Many prioniodinins have digyrate P elements.

There are few naturally occurring complete or partial apparatuses of prioniodinins, so positional homologies remain uncertain. For this reason, we retain the traditional element notation for this suborder, rather than applying the locational notation of Purnell *et al.* (2000).

Family PRIONIODINIDAE Bassler, 1925
Genus OULODUS Branson and Mehl, 1933c

?1981 *Guizhouprioniodus* Zhou *et al.*, 1981, p. 132.

Type species. Oulodus mediocris Branson and Mehl, 1933c; junior subjective synonym of *Oulodus serratus* (Stauffer, 1930), see Sweet and Schönlaub (1975, p. 45).

Diagnosis. Pa and Pb elements extensiform digyrate with prominent cusp; M makellate; Sa alate, normally without denticulate posterior process; Sb elements breviform digyrate; Sc elements dolabrate or bipennate. Process denticles of all elements tend to be stout, discrete, peg-like and separated from adjacent denticles by U-shaped spaces (updated in terminology from Sweet and Schönlaub 1975, p. 45).

Remarks. Silurian species referred to *Oulodus* possess bipennate Sc elements, unlike the dolabrate element in the type species. Zhang and Barnes (2002) referred some

---

EXPLANATION OF PLATE 18

Figs 1–6. *Oulodus* aff. *angullongensis* Bischoff, 1986. Xiangshuyuan Formation, Leijiatun Section, Shiqian County, Guizhou, Sample TT 813. 1, 149892, Pa element, posterior view. 2, 149893, Pb element, posterior view. 3, 149894, Sb element, posterior view. 4, 149895, Pa element oblique outer lateral view. 5, 149896, M element, posterior view. 6, 149897. Sc element, inner lateral view.

Figs 7–11. *Oulodus* aff. *panuarensis* Bischoff, 1986. Xiangshuyuan Formation, Leijiatun Section, Shiqian County, Guizhou, Sample Shiqian 3. 7, 149898, Pb element, posterior view. 8, 149899, M element, posterior view. 9, 149900, Sa element, posterior view. 10, 149901, Sb element, posterior view. 11, 149902, Sc element, inner lateral view.

Figs 12–18. *Oulodus shiqianensis* (Zhou *et al.*, 1981). Shenxuanyi Member, Xuanhe Section, Guangyuan County, Sichuan, Sample Xuanhe 2. 12, 149903, Pb element, posterior view. 13, 149904, Pa element, posterior view. 14, 149905, ?Sb element, posterior view. 15, 149906, Sa element, posterior view. 16, 149907, Sa element, lateral view. 17, 149908, Sc element, inner lateral view. 18, 14909, Sc element, inner lateral view.

Figs 19–21. *Oulodus shiqianensis* (Zhou *et al.*, 1981). Lower member, Xiushan Formation, Leijiatun Section, Shiqian County, Guizhou, Sample Shiqian 15. 19, 149910, Pa element, posterior view. 20, 149911, M? element, posterior view. 21, NHM X1133, Sa element, oral view.

All figures ×60.

PLATE 18

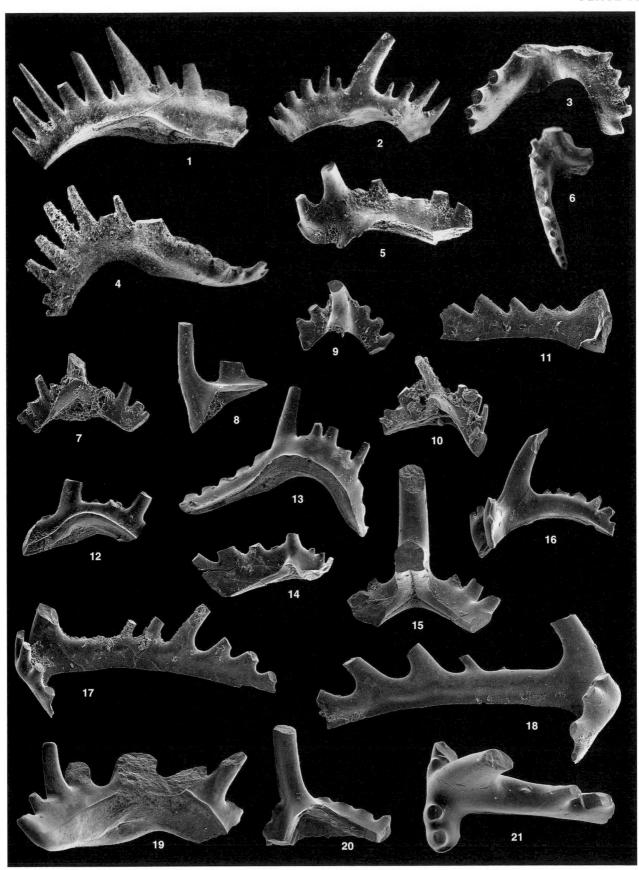

WANG and ALDRIDGE, *Oulodus*

Silurian species that had previously been equivocally assigned to *Oulodus* to their new genus *Rexroadus* (see below); *Rexroadus* elements differ from those of *Oulodus* in possessing compressed or slightly compressed cusp and denticles, especially on the P elements.

It is difficult to be certain of the nature of the single figured element of the type species of *Guizhouprioniodus*, *Guizhouprioniodus guizhouensis* Zhou et al, 1981, pl. 1, figs 13–14, as it is broken. However, its robust nature and aboral keel appear to be distinctive. Fordham (1991) provided a multielement concept of the genus, reconstructing an apparatus similar to that of *Oulodus*. Wang and Aldridge (1998) noted that many authors have a broad concept of *Oulodus*, which would probably encompass this model for the apparatus of *Guizhouprioniodus*; resolution of these relationships awaits detailed phylogenetic analysis.

In addition to the species described below, specimens of *Oulodus* occur in small numbers in many of the collections we have examined, but are commonly too few to be certain of species assignments. They are recorded simply as *Oulodus* sp. in the occurrence tables; a few are illustrated as *Oulodus* sp. A (Pl. 19, figs 16–17) and *Oulodus* sp. B (Pl. 19, fig. 18).

Oulodus aff. *angullongensis* Bischoff, 1986
Plate 18, figures 1–6

aff. 1986 *Oulodus angullongensis* Bischoff, p. 69, pl. 15, figs 23–35, pl. 16, figs 1–30 (multielement).

Material. Pa, 2; Pb, 1; M, 1; Sb, 1; Sc, 1; plus additional material from TT 813.

Description. Pa element arched, digyrate, with prominent cusp and two lateral processes of subequal length. Cusp subcircular to lenticular in cross-section, directed posteriorly and inwards. Inner, longer process extended downwards and twisted anteriorly, bearing up to eight peg-like, discrete denticles of variable height, smallest distally; proximal denticle may also be small. Outer process with up to six peg-like, discrete denticles, largest centrally. Basal cavity shallow and broad beneath cusp, narrowing slowly to process tips.

Pb element arched, digyrate, but less twisted than Pa, with more prominent cusp. Cusp subrounded to lenticular in cross-section, considerably larger than largest denticles, directed posteriorly and inwards. Processes of subequal length; inner process curved anteriorly, bearing at least four discrete, peg-like denticles. Outer process straighter, bearing about six discrete, peg-like denticles with circular or subcircular cross-sections; lower margin of process strongly arched. Basal cavity shallow, widest beneath cusp, tapering more rapidly under processes than on Pa.

M element digyrate. Cusp stout, circular in cross-section, thickening near its base, posteriorly curved. Inner lateral process short, slightly downturned, bearing one or two peg-like denticles. Outer lateral process much longer with up to six discrete denticles with lenticular cross-sections. Basal cavity shallow, widest beneath cusp, extending as broad groove along lower surface of outer lateral process.

Sb element digyrate. Arched, asymmetrical, cusp broken on available specimens. Lateral processes somewhat twisted with up to five discrete denticles, circular in cross-section. Basal cavity flared below cusp to produce a lip on posterior side.

Sc element bipennate, cusp posteriorly curved, subcircular in cross-section. Antero-lateral process relatively long for this element of *Oulodus*, directed strongly downwards, bearing up to nine discrete, slightly posteriorly curved, peg-like denticles. Posterior process broken on all specimens; long, straight, bearing more than four discrete, strongly posteriorly inclined denticles.

Remarks. Sa specimens have not been identified. The two P elements are very similar, but the cusp on the Pb element is more pronounced, and in this respect the element differs from its counterpart in the type

EXPLANATION OF PLATE 19

Figs 1–15. *Oulodus tripus* sp. nov. Lower member, Xiushan Formation, Leijiatun Section, Shiqian County, Guizhou, Sample Shiqian 14B. 1, 149912, Pa element, posterior view. 2, 149913, Pa element, posterior view. 3, 149914, Pb element, posterior view. 4, 149915, Pb element, posterior view. 5, 149916, Pb element, posterior view. 6, 149917, Pb element, posterior view. 7, 149918, M element, posterior view. 8, 149919, M element, posterior view. 9, 149920, M element, posterior view. 10, 149921, Sb element, posterior view. 11–12, 149922, Sa element, posterior view and close-up of basal cavity showing edges of lamellae (holotype). 13–14, 149923, Sb element, posterior and oral views. 15, 149924, Sa element, posterior view.
Figs 16–17. *Oulodus* sp. A. Leijiatun Formation, Leijiatun Section, Shiqian County, Guizhou, Sample Shiqian 7. 16, 149925, Pb element, posterior view. 17, 149926, Sc element, inner lateral view.
Fig. 18. *Oulodus* sp. B. Lower member, Xiushan Formation, Leijiatun Section, Shiqian County, Guizhou, Sample Shiqian 14B. 149927, Sb? element, oblique posterior view.
All figures ×60, except fig. 12, ×200.

PLATE 19

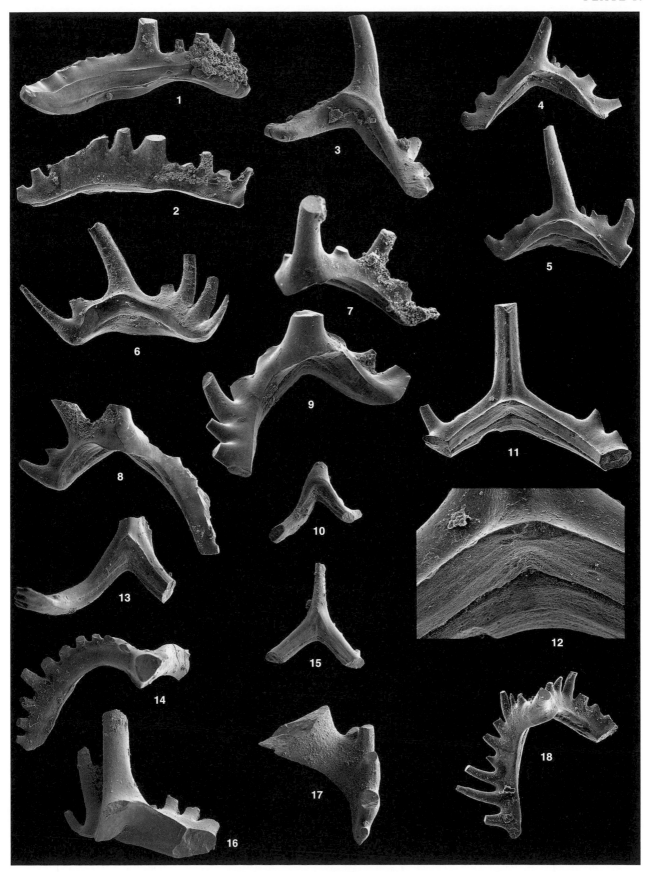

WANG and ALDRIDGE, *Oulodus*

collections from Australia (Bischoff 1986, pl. 15, figs 24, 31–35, if these are indeed Pb elements). The cavity is also narrower along the processes of the Pb element in the Chinese specimens. The long antero-lateral process of the Sc element in this material compares well with the same element in *O. angullongensis* (Bischoff 1986, pl. 16, figs 8–14); the specimen illustrated as *Lonchodina guidingensis* sp. nov. by Zhou *et al.* (1981, pl. 1, figs 22–23) has a similarly long antero-lateral process and may come from the same apparatus or a related one.

Occurrence. Xiangshuyuan Formation, Leijiatun section, Shiqian County, Guizhou (Sample TT 813).

Oulodus aff. *panuarensis* Bischoff, 1986
Plate 18, figures 7–11

aff. 1986 *Oulodus panuarensis* Bischoff, p. 75, pl. 19, figs 22–38 (multielement).

Material. Pb, 2; M, 1; Sa, 2; Sb, 2; Sc, 2.

Remarks. The M element of *O. panuarensis* is particularly characteristic, and Bischoff (1986, p. 75, pl. 19, fig. 25) designated a specimen of the M element as holotype of the species. A set of poorly preserved *Oulodus* specimens from sample Shiqian 3 include a single specimen of an M element (Pl. 18, fig. 8) that is very close in general form to that of *O. panuarensis*, with a strongly expanded basal cavity and a short outer lateral process. The Chinese specimen differs in having only a single, very compressed denticle on the outer lateral process, rather than up to four peg-like denticles. All elements of *O. panuarensis* have relatively large basal cavities (Bischoff 1986, p. 76; Zhang and Barnes 2002, p. 25), and the Chinese Sa and Sb specimens compare closely with those illustrated by Bischoff. The Chinese sample also includes two specimens of a Pb element with a large basal cavity (Pl. 18, fig. 7); Bischoff (1986) did not identify a Pb element. The only specimens of Sc elements in the Chinese sample are badly broken, but the more complete one (Pl. 18, fig. 11) shows more strongly compressed denticles on the posterior process than those illustrated by Bischoff (1986, pl. 19, figs 29–32). There is no Pa element in the small Chinese collection; the specimens assigned by Bischoff (1986, pl. 19, figs 22–24) to the Pa position are digyrate with large basal cavities and bear a very close resemblance to the Pb element of *Pseudolonchodina expansa* (Armstrong); a Pb specimen is the holotype of that species (Armstrong 1990, pl. 3, fig. 13). The other elements of *P. expansa* do not closely resemble their equivalents in *O. panuarensis*.

Occurrence. Xiangshuyuan Formation, Leijiatun section, Shiqian County, Guizhou (Sample Shiqian 3).

Oulodus shiqianensis (Zhou, Zhai and Xian, 1981)
Plate 18, figures 12–21

1981 *Hibbardella shiqianensis* Zhou *et al.*, p. 133, pl. 1, figs 20–21 (Sa element).
1983 *Hibbardella shiqianensis* Zhou, Zhai et Xian; Zhou and Zhai, p. 277, pl. 66, figs 7ab (Sa element).
v 1996 *Oulodus shiqianensis* (Zhou *et al.*); Wang and Aldridge, pl. 3 fig. 7 (Sa element).
v 2002 *Oulodus shiqianensis* (Zhou *et al.*); Aldridge and Wang, fig. 64G [copy of Wang and Aldridge 1996, pl. 3, fig. 7] (Sa element).

Diagnosis. Apparatus characterized by Sa element with widely arched, short lateral processes and a long, denticulate posterior process. Sc element with short antero-lateral process directed downwards and posteriorly and long posterior process with widely spaced denticles.

Material. Pa, 37; Pb, 66; M, 59; Sa, 42; Sb, 67; Sc, 108; plus additional material from TT samples.

Description. Pa element digyrate and strongly arched. Cusp fairly prominent, straight, inclined slightly posteriorly, with subcircular cross-section. Inner lateral process curved aborally and anteriorly, bearing up to seven peg-like denticles, discrete but quite closely spaced. Outer lateral process straighter, inclined anteriorly, bearing four to six denticles with lenticular cross-sections. Basal cavity shallow and wide beneath cusp and inner lateral process, tapering more rapidly beneath outer lateral process, except in larger, more robust specimens (see Pl. 18, fig. 19).

Pb element digyrate and gently arched. Cusp prominent, straight, inclined slightly posteriorly, with subcircular cross-section. Inner lateral process gently curved, broken on all specimens, but with at least four peg-like, discrete denticles. Outer lateral process relatively straight. Basal cavity shallow and wide beneath cusp and inner lateral process, tapering more rapidly beneath outer lateral process.

M element digyrate. Robust, with tall, slightly curved cusp with subcircular cross-section. Two denticulate processes, broken on all specimens; longer, outer lateral process bears discrete denticles with lenticular cross-section. Basal cavity wide and moderately deep beneath cusp, flaring posteriorly; continues as broad groove under outer lateral process, narrowing more rapidly beneath inner lateral process.

Sa element alate, with prominent cusp and well-developed posterior process. Cusp tall, curved, ovoid in transverse section. Lateral processes short, directed laterally or slightly antero-laterally, forming an arch of 140–160° in posterior view; each process bearing two or three discrete denticles, circular in cross-section. Posterior process long, straight or slightly downcurved, bearing up to six discrete, posteriorly inclined denticles of ovoid cross-section; proximal denticle may be considerably smaller than remainder. Basal cavity shallow; broad beneath posterior process, tapering beneath lateral processes.

?Sb element digyrate. Only equivocally identified. Appears to be broadly arched with one process bearing about four discrete denticles with lenticular cross-sections, other process with at least three smaller, peg-like denticles. Cusp with subcircular cross-section. Cavity flared posteriorly beneath cusp to form a lip, tapering gradually under processes.

Sc element bipennate with prominent curved cusp with subcircular to ovoid cross-section. Antero-lateral process directed strongly downwards and a little posteriorly, bearing three to five discrete denticles with subcircular to ovoid cross-sections, curving very slightly posteriorly. Posterior process long, very gently downcurved; there is a long gap between cusp and proximal denticle, which may be tiny. Denticles up to seven in number, all widely spaced, increasing in size distally, ovoid in transverse section and curving gently posteriorly. Basal cavity a shallow groove beneath processes, not widening beneath cusp.

Remarks. The Sa element of this species is characteristic, bearing a well-developed, denticulate posterior process. Species of *Oulodus* normally lack this process, and *O. shiqianensis* might be better assigned to a separate genus; however, all the other elements of the apparatus are distinctly *Oulodus*-like and, in the absence of phylogenetic evidence, we retain it in *Oulodus* here. As with some other species of *Oulodus*, the Pa and Pb elements are difficult to differentiate, and the discrimination of these may be wrong; a natural assemblage would help to solve this problem.

The specimens from the Welsh Borderland of England referred to *Hibbardella* sp. nov. by Aldridge (1972, p. 182., pl. 6, figs 14–15) are very similar to the Sa element of *O. shiqianensis* and were included in synonymy with *H. shiqianensis* by Zhou *et al.* (1981). However, the associated Sc specimens (Aldridge 1972, pl. 8, figs 10, 13, 16) have antero-lateral processes that are perpendicular to the posterior process, pointing directly downwards; they also possess closely spaced denticles on the posterior process. The English specimens probably belong to a separate species that may well be related to *O. shiqianensis*.

Occurrence. Leijiatun Formation and lower and upper members, Xiushan Formation, Leijiatun section, Shiqian County, Guizhou; Yangpowan Member, Ningqiang Formation, Yushitan section, Ningqiang, Shaanxi; Shenxuanyi Member, Xuanhe section, Guangyuan, Sichuan.

Oulodus tripus sp. nov.
Plate 19, figures 1–15

Derivation of name. L., *tripus*, a three-legged stand; in reference to the three-rayed shape of the characteristic Sa element.

Holotype. Specimen NIGPAS 149922 (Pl. 19, figs 11–12); Sa element.

Type locality and horizon. Leijiatun section, Shiqian County, Guizhou; lower member, Xiushan Formation, sample Shiqian 14B.

Diagnosis. All elements with broad, shallow, sometimes recessive basal cavities. Pa element nearly straight, with erect cusp and somewhat twisted processes; Pb element variably arched. Sa and Sb elements with tall cusps with triangular cross-sections; Sa element lacks posterior process. Distal portions of processes on Sb element strongly flexed.

Material. Pa, 9; Pb, 17; M, 12; Sa, 19; Sb, 13.

Description. Pa element digyrate, nearly straight in lateral view. Cusp prominent, erect, with ovoid cross-section. Inner lateral process straight or slightly curved, bearing four or five discrete, peg-like denticles of subequal size; gap between cusp and first denticle may be relatively large. Outer lateral process directed slightly aborally, twisted slightly clockwise, bearing up to seven compressed and partially fused denticles, which may become discrete and peg-like distally. Basal cavity shallow and wide beneath entire element, recessive along entire length (Pl. 19, fig. 1) or beneath outer lateral process (Pl. 19, fig. 2).

Pb element digyrate and gently to strongly arched. Cusp very prominent, inclined gently posteriorly and curved slightly inwards, with subcircular cross-section. Inner lateral process with a straight base, twisted strongly clockwise, with two to four peg-like, discrete denticles. Outer lateral process straight, only slightly twisted, with three or four discrete denticles, ovoid in transverse section. Basal cavity shallow and wide beneath cusp and processes, becoming recessive in more robust specimens (Pl. 19, fig. 6).

M element digyrate; robust, with tall, slightly curved cusp with subcircular cross-section. Outer lateral process bears up to six discrete denticles with lenticular cross-section. Inner lateral process directed less strongly aborally, with up to three discrete, peg-like denticles. Base of cusp characterized by a broad posterior flare, which hangs over basal cavity. Basal cavity wide and shallow, tapering gradually beneath longer process, but broader, becoming recessive, under shorter process (Pl. 19, figs 8–9).

Sa element alate, with prominent cusp and no posterior process. Cusp tall, straight, triangular in transverse section, with a broad costa on posterior face. Lateral processes long, broken on all specimens, directed laterally and inclined anteriorly, forming an arch of 120–140° in posterior view; each process bearing at least four small, discrete, widely spaced denticles, ovoid in cross-section. Basal cavity shallow and very broad, tapering only gradually beneath lateral processes.

Sb element digyrate, with long processes forming an asymmetrical arch of about 90°. Cusp tall, straight, triangular in cross-section. Lateral processes commonly broken, but are long and strongly flexed so that distal tip is in a plane perpendicular to the proximal end; each bears up to ten discrete denticles of subequal size and ovoid cross-section. Basal cavity shallow and broad, tapering only gradually beneath lateral processes.

Sc element not recognized.

Remarks. This is a distinctive apparatus, with all elements having broad basal cavities; the Sa and Sb elements are particularly characteristic. The Sc elements have not been identified; specimens may be included in those assigned to *Ctenognathodus*? *qiannanensis*, which occurs in large numbers in the same sample. The Pa element bears some similarities to that of *Rexroadus* species, in the presence of compressed, partially fused denticles on one process and discrete denticles on the other. The species is assigned to *Ouldodus* rather than *Rexroadus* on the basis of the ovoid cross-section of the cusp of the Pa element and because of the peg-like denticles on the other elements. The close morphological similarities in the Pa elements, however, draw attention to an orientational problem in this group of taxa: the processes on *Ouludus* elements are regarded as lateral and those of *Rexroadus* are considered to be anterior and posterior. It is highly likely that these processes are homologous, and this will have to be taken into consideration in future phylogenetic analyses.

Occurrence. Lower member, Xiushan Formation, Leijiatun section, Shiqian County, Guizhou; lower part of Shenxuanyi Member, Xuanhe section, Guangyuan, Sichuan.

Genus PSEUDOLONCHODINA Zhou, Zhai and Xian, 1981

1981 *Neoplectospathodus*, Zhou *et al.*, p. 134.
1985 *Aspelundina* Savage, p. 725.

Type species. Pseudolonchodina irregularis Zhou *et al.*, 1981, p. 136, by monotypy.

Emended diagnosis (based on emended diagnosis for *Aspelundia* given by Armstrong, 1990, p. 49). Pa and Pb elements digyrate with twisted processes. M dolabrate, Sa alate with posteriorly extended basal cavity, Sb tertiopedate, Sc bipennate. All elements highly compressed.

Remarks. Walliser (1964) described two new taxa, *Lonchodina fluegeli* and *Neoprioniodus planus*, that have subsequently been considered to be from a single prioniodinid apparatus. Sweet and Schönlaub (1975) included them in the multielement *Ozarkodina plana* (Walliser), but this reconstruction appears to have included elements from more than one species (see Aldridge 1979). Aldridge (1979) considered Walliser's specimens to be the Pa, Pb and M elements of multielement *Ouludus*? *fluegeli* (Walliser), preferring not to use the species name *planus*, as the M element appears to be indistinguishable from that of *Ozarkodina polinclinata* (Nicoll and Rexroad), as reconstructed by Cooper (1977. p. 1058), rendering the apparatus of Walliser's original specimens uncertain. Savage (1985), however, considered that Walliser's two taxa belonged to two different multielement species, *Ouludus fluegeli* and *Pandorinellina plana*. He also erected an additional species, *Aspelundia capensis*, type species of his new genus *Aspelundia*, to accommodate some morphologically similar elements in which an additional process is developed (Savage 1985, p. 725, fig. 19). Bischoff (1986) reconstructed an apparatus similar, but not identical, to that proposed by Aldridge (1979) and applied the name *Ouludus planus*. Armstrong (1990)

EXPLANATION OF PLATE 20

Fig. 1. ?*Pseudolonchodina expansa* (Armstrong, 1990). Leijiatun Formation, Leijiatun Section, Shiqian County, Guizhou, Sample Shiqian 7. 149928, Pb element, posterior view.

Figs 2–9. *Pseudolonchodina fluegeli* (Walliser, 1964). Xiangshuyuan Formation, Leijiatun Section, Shiqian County, Guizhou, Sample Shiqian 5. 2, 149929, Pa element, posterior? view. 3, 149930, Pb element, posterior view. 4, 149931, M element, inner lateral view. 5, 149932, Sc element, inner lateral view. 6, 149933, Sb element, posterior view. 7, 149934, Sa element, lateral view. 8–9, 149935, Sa element, posterior and anterior views.

Figs 10–16. *Pseudolonchodina fluegeli* (Walliser, 1964). Upper member, Xiushan Formation, Leijiatun Section, Shiqian County, Guizhou, Sample Shiqian 18. 10–11, 149936, Pa element. 12, 149937, Pb element, posterior view. 13, 149938, Pb element, posterior view. 14, 149939, M element, inner lateral view. 15, 149940, Sc element, inner lateral view. 16, 149941, Sc element, inner lateral view.

Figs 17–19. *Pseudolonchodina fluegeli* (Walliser, 1964). Upper member, Xiushan Formation, Leijiatun Section, Shiqian County, Guizhou, Sample Shiqian 17. 17–18, 149942, Sa element, oral and lateral views. 19, 149943, Sb element, posterior view.

Figs 20–24. *Pseudolonchodina* sp. Kuanyinchiao Bed, Leijiatun Section, Shiqian County, Guizhou, Sample Shiqian-1. 20, 149944, Pb element, posterior view. 21, 149945, Sb? element, lateral view. 22, 149946, Pb element, posterior view. 23, 149947, Sc element, inner lateral view.

All figures ×80.

PLATE 20

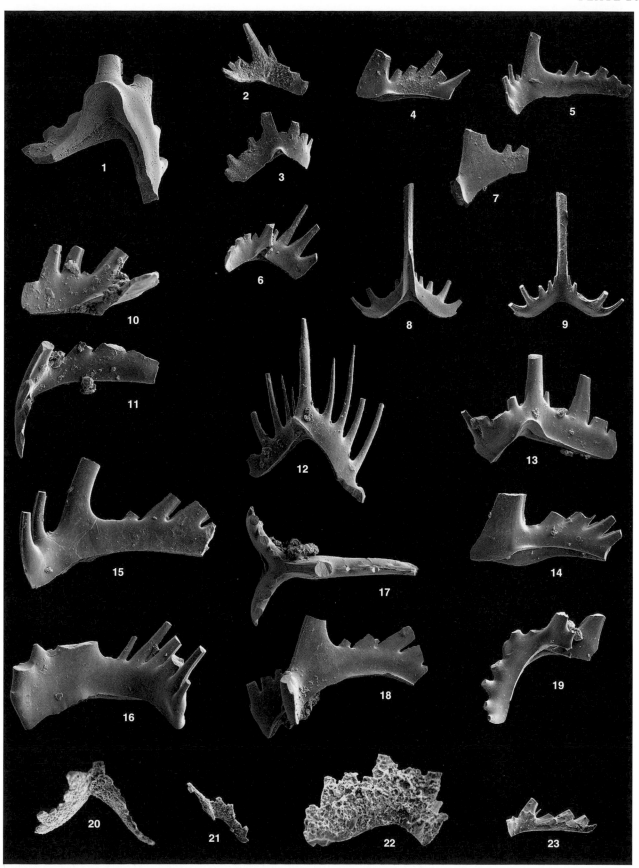

WANG and ALDRIDGE, *Pseudolonchodina*

accepted the reconstruction published by Aldridge (1979), and also considered the specific name *planus* to be a *nomen dubium*. He did not place the species *fluegeli* in *Oulodus*, but transferred it to *Aspelundia*.

This chequered taxonomic and nomenclatural history is further complicated by the description of two single-element taxa by Zhou *et al.* (1981, pp. 134, 136, pl. 1, figs 32–33, 45–47): *Neoplectospathodus luomianensis* and *Pseudolonchodina irregularis*. These elements have three processes, and compare closely with elements assigned by Savage (1985) to *Aspelundia capensis*. Simpson and Talent (1995) provided a full discussion of these similarities, and considered *Pseudolonchodina* and *Aspelundia* to be synonymous, with *Neoplectospathodus* Zhou *et al.* also tentatively synonymous. *Neoplectospathodus* Zhou *et al.*, 1981, is a junior homonym of *Neoplectospathodus* Kozur and Mostler, 1970, a Triassic conodont, and hence is not an available name for this multielement taxon. Fordham (1991) included both *Pseudolonchodina* and *Aspelundia* as junior synonyms of *Diadelognathus* Nicoll and Rexroad, 1968, but elements of that genus bear separated peg-like denticles more akin to those of *Oulodus*.

There is some uncertainty about the positions of some of the elements within the apparatus of *Pseudolonchodina*, especially the relative positions of the P elements. We note the inclusion in the apparatus by Armstrong (1990, pl. 3, figs 8–9) of an element with a single lateral process, termed Sd by him. We have not found specimens of this morphology in the Chinese collections, but accept that this form of element is present in some *Pseudolonchodina* apparatuses, perhaps occupying an S_1 or an S_2 position.

?*Pseudolonchodina expansa* (Armstrong, 1990)
Plate 20, figure 1

* v ?1990 *Aspelundia expansa* Armstrong, p. 50, pl. 3, figs 13–20 (multielement).

Remarks. A single specimen of a breviform digyrate element, morphologically close to that assigned to the Pb position of *P. expansa* by Armstrong (1990) has been recovered from the Leijiatun Formation (sample Shiqian 7); the principal difference is that the cavity is even more expanded than in the specimen figured by Armstrong. In this respect, it closely resembles the holotype 'Pa element' of *Oulodus australis* Bischoff (Bischoff 1986, p. 72, pl. 16, fig. 31), although the disposition of the processes is a little different. Associated broken specimens (Pl. 19, figs 16–17) lack the characteristic morphology of *Pseudolonchodina* elements, and are referred to *Oulodus* sp. A, so the assignment of this breviform digyrate element must be uncertain.

Occurrence. Leijiatun Formation, Leijiatun section, Shiqian County, Guizhou.

Pseudolonchodina fluegeli (Walliser, 1964)
Plate 20, figures 2–19

*1964 *Lonchodina fluegeli* Walliser, p. 44, pl. 6, fig. 4; pl. 32, figs 22–24 (Pa, Pb elements).

1964 *Neoprioniodus planus* Walliser, p. 51, pl. 4, fig. 10; pl. 6, fig. 3; pl. 29, figs 12–13, 15 (M element).

1964 ?*Roundya trichonodelloides* Walliser, p. 72, pl. 6, fig. 2; pl. 31, figs 22–25 (Sa element).

1983 *Hibbardella trichonodelloides* (Walliser, 1964); Zhou and Zhai, p. 277, pl. 66, fig. 8a–b (Sa element).

1983 *Lonchodina detorta* Walliser, 1964; Zhou and Zhai, p. 280, pl. 66, figs 12–13 (Pb element).

1983 *Lonchodina* cf. *fluegeri* Walliser, 1964 [*sic*]; Zhou and Zhai, p. 280, pl. 66, fig. 14a–b (Pa element).

1983 *Cyrtoniodus planus* (Walliser, 1964); Zhou and Zhai, p. 271, pl. 66, figs 20–21 (M element).

1986 *Oulodus planus planus* (Walliser, 1964); Bischoff, p. 80, pl. 19, figs 39–41; pl. 20, figs 1–7, 17–44; pl. 21, figs 1–12 (multielement).

1987 *Lonchodina fluegeli* Walliser sf.; Ni, p. 416, pl. 61, figs 2, 34 (Pb element).

?1987 *Neoprioniodus planus* Walliser sf.; Ni, p. 419, pl. 61, figs 15, 25 (M element).

1989 *Lonchodina fluegeri* Walliser [*sic*]; Yu *in* Jin *et al.*, pl. 2, fig. 5 (Pa element).

p 1989 *Lonchodina detorta* Walliser, 1964; Yu *in* Jin *et al.*, p. 103, pl. 7, fig. 16 only (Pb element).

?1989 *Plectospathodus extensus* Rhodes; Yu *in* Jin *et al.*, pl. 2, fig. 19; ?pl. 4, fig. 10 (Sb element).

?1989 *Hindeodella equidentata* Rhodes; Yu *in* Jin *et al.*, pl. 4, fig. 3; pl. 6, fig. 4 (Sc element).

?1989 *Hibbardella trichonodelloides* (Walliser); Yu *in* Jin *et al.*, pl. 4, figs 8, 14 (Sa element).

?1989 *Neoprioniodus planus* Walliser; Yu *in* Jin *et al.*, p. 103, pl. 5, fig. 14; pl. 7, fig. 17 (M element).

v 1990 *Aspelundia fluegeli* (Walliser, 1964); Armstrong, p. 53, pl. 3, figs 1–9, 11–12 (multielement, with synonymy to 1987).

1991a *Aspelundia fluegeli* (Walliser); McCracken, p. 73, pl. 1, figs 5–6, 10, 13–14, 17, 19–24, 26–27, 31–32, pl. 2, figs 3–4, 7–8, 10–11, 13–15, 17–25, 29, 32 (multielement, with synonymy to 1990).

1995 *Pseudolonchodina fluegeli* (Walliser, 1964); Simpson and Talent, p. 112, pl. 1, fig. 4 (?Sb element, with synonymy to 1994).

v 1996 *Pseudolonchodina fluegeli* (Walliser); Wang and Aldridge, pl. 3, fig. 6 (Pa element).

1996 *Aspelundia fluegeli* (Walliser, 1964); Girard and Weyant, p. 56, pl. 2, figs 6a–b, 7, 8a–b (Pb, M, Sa elements).

1997 *Aspelundia fluegeli* (Walliser); Dumoulin *et al.*, fig. 4EE–GG (Pb. Pa, Pb elements).

1999 *Pseudolonchodina fluegeli* (Walliser, 1964); Simpson, p. 189, pl. 2, figs 1–2 (Sa, Sc elements).

v 2002 *Pseudolonchodina fluegeli* (Walliser); Aldridge and Wang, fig. 64F (copy of Wang and Aldridge 1996, pl. 3, fig. 6) (Pa element).

Material. Pa, 167; Pb, 110; M, 94; Sa, 85; Sb, 64; Sc, 241; plus additional material from TT samples.

Remarks. This taxon was fully described and illustrated as *O. planus planus* by Bischoff (1986); we follow his reconstruction here. Stratigraphically older collections conform well to his diagnosis, for example the specimens illustrated from the Xiangshuyuan Formation in the Leijiatun section (sample Shiqian 5, Pl. 20, figs 2–9). Stratigraphically younger populations from the same section (e.g. from samples Shiqian 17 and 18, upper member of Xiushan Formation, Pl. 20, figs 10–19) show some of the characteristics that Bischoff (1986, p. 84) used to distinguish the subspecies *O. planus borenorensis* Bischoff. These characters include the presence of Sc specimens with the posterior portion of the posterior process bent strongly inwards (Pl. 20, fig. 16), and of an Sb element with the shorter process bearing a large denticle with one or two smaller denticles (Pl. 20, fig. 19). However, other diagnostic or characteristic features of this subspecies are not evident; for example, the anterior process of the Sc element of *borenorensis* is long and high with 8–11 denticles, and the basal cavity of Pa and Pb elements is strongly reduced. In general, elements referred to *P. fluegeli* are very variable, and it may well be possible to distinguish several subspecies/species; for example, the posterior process of the Sa element may be adenticulate (Pl. 20, figs 8–9) or long and denticulate (Pl. 20, figs 17–18); both morphotypes were included in *O. p. borenorensis* by Bischoff (1986, p. 85). As the full apparatus of Walliser's type material is unknown, all these variants are currently best accommodated in a broad concept of *P. fluegeli*.

Some small specimens from the Kuanyinchiao Bed of the Leijiatun section (sample Shiqian −1; Pl. 20, figs 20–23), regarded as latest Ordovician in age, represent a species of *Pseudolonchodina*. The Pb element resembles that of *P. fluegeli*, but has a more restricted cavity; the collection is small and not all elements of the apparatus are present, so these specimens are identified simply as *Pseudolonchodina* sp.

Occurrence. Xiangshuyuan, Leijiatun and Xiushan formations, Leijiatun section, Guizhou; Yangpowan and Shenxuanyi members, Ningqiang Formation, Yushitan section, Ningqiang, Shaanxi; Shenxuanyi Member, Xuanhe section, Guangyuan, Sichuan.

Pseudolonchodina sp. nov. A
Plate 21, figs 1–6

Material. Pa, 1; Pb, 3; M, 3; Sa, 6; Sb, 3; Sc, 17,

Remarks. A suite of elements from sample Xuanhe 4 displays the general characters of *Pseudolonchodina*, but differs in several respects from *P. fluegeli*; some of the specimens that appear to belong to this apparatus show some similarity to corresponding elements in *Oulodus*. Pa element broken, compressed digyrate, with a tall, laterally compressed cusp; each process twisted with respect to the other. Pb element digyrate, broadly arched with a more extended basal cavity than is typical for *Pseudolonchodina*; both processes flexed. M element dolabrate, planate, with a relatively low, laterally compressed cusp; lower anterior margin of cusp extended slightly downwards into a short, rounded anticusp; posterior process with confluent denticles, decreasing in height distally. Sa element alate, with short lateral processes and long posterior process, which bears at least four widely spaced denticles, ovoid in cross-section and increasing in size distally. Sb element digyrate, with a flared cavity beneath the cusp. Sc element bipennate, with anterior process short and directed downwards; posterior process straight with at least four denticles, ovoid in cross-section and separated by U-shaped spaces.

Occurrence. Shenxuanyi Member, Xuanhe section, Guangyuan County, Sichuan (sample Xuanhe 4).

Genus REXROADUS Zhang and Barnes, 2002

Type species. *Oulodus? nathani* McCracken and Barnes, 1981.

Remarks. This genus was separated from *Oulodus* by Zhang and Barnes (2002, p. 35) on the basis of the distinctive Pa element, which is blade-like with partially fused denticles on the straight anterior process and a posterior process with discrete denticles that are slightly directed laterally. All elements have a cusp and denticles that are compressed or slightly compressed. A few specimens in the Chinese collections that correspond to the diagnosis (Pl. 21, figs 13–16) are designated as *Rexroadus* spp A and B, as the complete apparatuses have not been identified; the Pa element here referred to *Rexroadus* sp. A (Pl. 21, fig. 13) has some similarities to *R. nathani* (McCracken and Barnes, 1981, pl. 6, figs 31–32), but the anterior process is lower. A set of elements from sample Shiqian 17 may represent a *Rexroadus* apparatus, and they are identified here as *Rexroadus* aff. *kentuckyensis*.

Rexroadus aff. *kentuckyensis* (Branson and Branson, 1947)
Plate 21, figures 7–12

aff. *1947 *Ligonodina kentuckyensis* Branson and Branson,
    p. 555, pl. 82, figs 28, 35 (Sc element).
aff. 1981 *Oulodus? kentuckyensis* (Branson and Branson);
    McCracken and Barnes, p. 80, pl. 6, figs 1–20
    (multielement).
aff. 2002 *Rexroadus kentuckyensis* (Branson and Branson,
    1947); Zhang and Barnes, p. 35, figs 10.37–10.48
    (multielement).

Material. Pa, 4; ?Pb, 4; ?M, 13; ?Sa, 17; ?Sb, 21; ?Sc, 31.

Description. Pa element carminate to slightly digyrate, bowed and gently arched; specimens mostly badly broken. Anterior process of best-preserved specimen with about six denticles of uneven size, largest in central part of process, partly fused and strongly laterally compressed. Cusp broken away, but basal section highly compressed; posterior process slightly twisted, bearing three discrete denticles with lenticular transverse sections, separated by U-shaped spaces and inclined posteriorly and slightly laterally. Basal cavity a shallow groove, widening beneath cusp and adjacent denticle on each process.

?Pb element angulate, arched and bowed. Cusp very prominent with extended keels on posterior and, particularly, anterior edges; lenticular in transverse section. Anterior process with five laterally compressed denticles, fused basally and decreasing in size distally. Posterior process broken on all specimens, with at least three, discrete, peg-like denticles. Basal cavity flaring gently inwards beneath cusp and continuing as a shallow groove beneath processes.

?M element robust, makellate. Cusp very prominent, tall, lateral edges extended into broad keels; transverse section lenticular with pinched extremities. Inner lateral process directed aborally with up to four small, discrete denticles. Outer lateral process curving downwards, bearing up to nine closely spaced, somewhat compressed denticles. Basal cavity flaring inwards below cusp and continuing as a broad groove beneath outer lateral process.

?Sa element alate, lacking posterior process, but with strongly extended posterior lip to basal cavity. Cusp erect, prominent, with gently convex anterior face and posterior face that is flat at the margins but strongly convex axially. Lateral processes directed gently downwards and curving posteriorly, bearing up to six closely spaced denticles, lenticular in cross-section.

?Sb element digyrate, with slightly twisted cusp of subrounded transverse section with keeled lateral edges. One lateral process directed downwards and posteriorly, with about four compressed discrete denticles, with the central ones being very tall. Other lateral process directed more strongly downwards, bearing discrete denticles of subequal size and with subcircular transverse sections. Basal cavity forms a posterior lip beneath the cusp and extends as a shallow groove beneath the processes.

?Sc element bipennate, with a prominent, posteriorly curved cusp with an inner antero-lateral costa. Anterior process directed aborally, bearing three or four discrete denticles of subequal size and with subcircular transverse sections. Posterior process gently curved, with up to six discrete, posteriorly inclined denticles of lenticular cross-section, increasing in size distally. Basal cavity not flared below cusp, continuing as a groove beneath processes.

Remarks. The Pb, M and S elements all have similar denticulation and a tendency to develop keels on the cusp. They are all likely to be from a single apparatus, but bear resemblances to elements from *Wurmiella*. This may indicate a close relationship between *Rexroadus* and *Wurmiella*, or may mean that elements from two different species are associated in the sample and that the Pa element is from a separate taxon.

Occurrence. Upper member, Xiushan Formation, Leijiatun section, Shiqian County, Guizhou (sample Shiqian 17).

Suborder OZARKODININA Dzik, 1976

Diagnosis. Ozarkodinids with carminate, stellate or segminate P$_1$ elements. P$_2$ elements with 'anterior' and 'posterior' processes, commonly angulate.

EXPLANATION OF PLATE 21

Figs 1–6. *Pseudolonchodina* sp. nov. A. Shenxuanyi Member, Xuanhe Section, Guangyuan County, Sichuan, Sample Xuanhe 4. 1, 149948, Pa element. 2, 149949, Pb element, posterior view. 3, 149950, M element, inner lateral view. 4, 149951, Sa element, lateral view. 5, 149952, Sb element, posterior view. 6, 149953, Sc element, inner lateral view.

Figs 7–12. *Rexroadus* aff. *kentuckyensis* (Branson and Branson, 1947). Upper member, Xiushan Formation, Leijiatun Section, Shiqian County, Guizhou, Sample Shiqian 17. 7, 149954, Pa element, inner lateral view. 8, 149955, ?Pb element, inner lateral view. 9, 149956, ?M element, posterior view. 10, 149957, ?Sa element, posterior view. 11, 149958, ?Sb element, oblique lateral view. 12, 149959, ?Sc element, inner lateral view.

Fig. 13. *Rexroadus* sp. A. Xiangshuyuan Formation, Leijiatun Section, Shiqian County, Guizhou, Sample Shiqian 5. 149960, Pa element, lateral view.

Figs 14–16. *Rexroadus* sp. B. Leijiatun Formation, Leijiatun Section, Shiqian County, Guizhou, Sample Shiqian 7. 14, 149961, Pa element, inner lateral view. 15, 149962, ?Sb element, posterior view. 16, 149963, ?Sc element, inner lateral view.

Figs 1–6, 13–16, ×80; figs 7–12, ×60.

PLATE 21

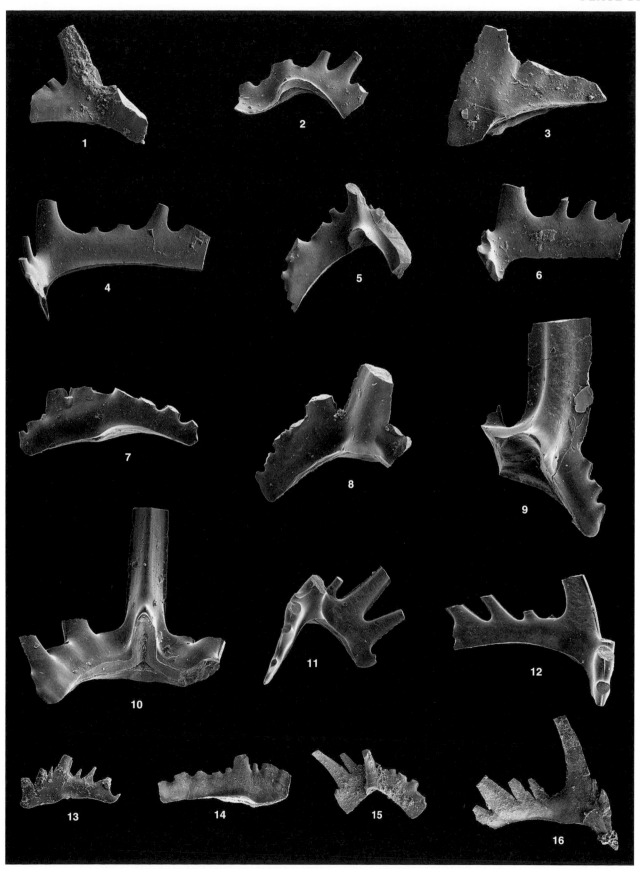

WANG and ALDRIDGE, *Pseudolonchodina, Rexroadus*

Remarks. Apparatus structure and element homologies are well understood within the ozarkodinins, so the locational nomenclature of Purnell *et al.* (2000) is applied.

Family SPATHOGNATHODONTIDAE Hass, 1959

Diagnosis. Ozarkodinins with bladelike carminate and angulate P elements (after Sweet 1988, p. 91).

Remarks. This grouping, as envisaged by Sweet (1988, pp. 91, 187) is undoubtedly polyphyletic (see Donoghue *et al.* 2008), but we retain it here pending a thorough cladistic analysis of the ozarkodinins. The genus *Ozarkodina* will be a member of any more narrowly diagnosed family.

Genus OZARKODINA Branson and Mehl, 1933a

1933a *Ozarkodina* Branson and Mehl, p. 51.
1933a *Plectospathodus* Branson and Mehl, p. 47.
1933a *Spathodus* Branson and Mehl, p. 46.
1941 *Spathognathodus* Branson and Mehl, p. 98.
1983 *Paraspathognathodus* Zhou, Zhai and Xian, p. 292.

Diagnosis. P_1 element carminate, P_2 angulate, M dolabrate, makellate or bipennate, S_{3-4} bipennate, S_{1-2} digyrate, S_0 alate (amended from Barrick and Klapper 1976, p. 78).

Type species. *Ozarkodina typica* Branson and Mehl, 1933a, p. 51.

Remarks. The genus *Ozarkodina* has a very generalized morphology and currently accommodates a large number of species, which may represent several clades. Several species currently assigned to other genera may also belong to these clades and should carry the same generic name, but the circumscription of *Ozarkodina* awaits a full cladistic analysis.

Ozarkodina broenlundi Aldridge, 1979
Plate 22, figure 1

*v 1979 *Ozarkodina broenlundi* Aldridge, p. 16, pl. 1 figs 18–25 (multielement).
v 1990 *Ozarkodina broenlundi* Aldridge, 1979; Armstrong, p. 88, pl. 12, figs 4–12 (multielement).
v 1996 *Ozarkodina broenlundi* Aldridge; Wang and Aldridge, pl. 3, fig. 8 (P_1 element).
v 2002 *Ozarkodina broenlundi* Aldridge; Aldridge and Wang, fig. 64H [copy of Wang and Aldridge 1996, pl. 3, fig. 8] (P_1 element).

Diagnosis. All elements with shallow, restricted basal cavities. P_1 element carminate with a prominent cusp and a large denticle immediately anterior to the cusp; denticles at anterior end of blade high. P_2 element with closely packed denticles and an undulose aboral edge. (modified after Aldridge 1979, p. 16).

Material. P_1, 32; ?P_2, 2; ?S_{1-2}, 1; ?S_{3-4}, 2.

Remarks. *Ozarkodina broenlundi* only occurs in small numbers in the Chinese collections and is always associated with other species of *Ozarkodina*, so only the P_1 element has been conclusively identified. There is a similarity to the P_1 element of *O. waugoolaensis*, which also has a prominent denticle anterior to the cusp but lacks the crest of high denticles at the anterior end of the blade.

EXPLANATION OF PLATE 22

Fig. 1. *Ozarkodina broenlundi* Aldridge, 1979. Wangjiawan Formation, Yushitan Section, Ninqiang County, Shaanxi, Sample Ningqiang 6. 149964, P_1 element, lateral view.

Figs 2–5. *Ozarkodina* aff. *cadiaensis* Bischoff, 1986. Shenxuanyi Member, Xuanhe Section, Guangyuan County, Sichuan, Sample Xuanhe 1. 2, 149965, P_1 element, oral view. 3, 149966, P_1 element, lateral view. 4, 149967, P_2 element, inner lateral view. 5, 149968, P_2 element, inner lateral view.

Figs 6–12. *Ozarkodina crispa* (Walliser, 1964), β morphotype. Sashuiyan Formation, Yuanyangyan-Maliuqiao Section, E. Erlangshan district, Sichuan, Sample TT 1056. 6, 149969, P_2 element, inner lateral view. 7, 149970, S_0 element, posterior view. 8, 149971, P_2 element, inner lateral view. 9, 149972, P_1 element, oral view. 10, 149973, P_1 element, oral view. 11, 149974, S_{3-4} element, inner lateral view. 12, 149975, S_{3-4} element, inner lateral view.

Figs 13–19. *Ozarkodina crispa* (Walliser, 1964), γ morphotype. Sashuiyan Formation, Yuanyangyan-Maliuqiao Section, E. Erlangshan district, Sichuan, Sample TT 1057. 13, 149976, P_1 element, oral view. 14, 149977, P_1 element, oral view. 15, 149978, S_{1-2} element, posterior view. 16, 149979, M element, inner lateral? view. 17, 149980, P_2 element, inner lateral view. 18, 149981, S_{1-2} element, posterior view. 19, 149982, S_{1-2} element, posterior view.

All figures ×80.

PLATE 22

WANG and ALDRIDGE, *Ozarkodina*

Occurrence. Lower member, Xiushan Formation, Leijiatun section, Shiqian County, Guizhou; Wangjiawan Formation, Yushitan section, Ningqiang, Shaanxi.

Ozarkodina aff. *cadiaensis* Bischoff, 1986
Plate 22, figures 2–5

aff. 1986 *Ozarkodina cadiaensis* Bischoff, p. 132, pl. 24,
   figs 11–27, 30 (multielement).
?p 1989 *Ozarkodina ludingensis* Yu *in* Jin *et al.*, p. 104,
   pl. 6, fig. 9 [*non* pl. 4, figs 15–16] (P$_2$ element).

Material. P$_1$, 5; P$_2$, 2; plus additional material from TT samples.

Remarks. The diagnosis of *O. cadiaensis* given by Bischoff (1986, p. 132) relates only to the P$_1$ element. He noted even denticulation interrupted in the central region by somewhat higher 'main and secondary' cusps enclosing a V-shaped gap, and a very short basal cavity, restricted to the area under the main cusp and with laterally strongly expanded lips. The specimens from south China have a similar basal cavity, but the lips are directed perpendicular to the blade, not slightly anteriorly as is characteristic of *O. cadiaensis*. The denticulation also differs, being even throughout and lacking a clearly defined larger cusp or adjacent denticle. The P$_2$ element associated with the P$_1$ specimens from south China has less crowded denticles than shown by the specimens illustrated by Bischoff (1986, pl. 24, figs 22–27), and the posterior process is much shorter and inwardly curved. Only small numbers of the P$_1$ and P$_2$ elements are present in the Chinese collections, and we have not recognized the M and S elements. We have, therefore, not proposed a new species name for these specimens.

Occurrence. Lower member, Xiushan Formation, Leijiatun section, Shiqian County, Guizhou (sample Shiqian 16); Shenxuanyi Member, Xuanhe section, Guangyuan County, Sichuan (samples Xuanhe 1, TT 463).

Ozarkodina crispa (Walliser, 1964)
Plate 22, figures 6–19

*1964 *Spathognathodus crispus* Walliser, p. 74, pl.9, fig.3;
   pl. 21, figs 7–13 (P$_1$ element).
1987 *Spathognathodus crispus* Walliser 1964; An, p. 202,
   pl. 34, figs 20–21 (P$_1$ element).
p 1989 *Spathognathodus crispus* Walliser, 1964; Yu *in* Jin
   et al., p. 107, pl. 8, figs 2a–b, 4a–b, 9a–b, 12a–c only
   [*non* fig. 11a–b indet.] (P$_1$ element).
1989 *Ozarkodina denckmanni* Ziegler; Yu *in* Jin *et al.*,
   p. 104, pl. 8, figs 7, 10 (P$_2$ element).

non-1992 *Spathognathodus crispus* Walliser, 1964; Qian *in*
   Jin *et al.* p. 61, pl. 3, figs 13, 17 (= *O. snajdri*
   (Walliser)).
v 1998 *Ozarkodina crispa* (Walliser, 1964); Viira and
   Aldridge, 39, pl. 2, figs 1–22; pl. 3, figs 1–11
   (P$_1$ element; with synonymy to 1995).
v 2001 *Ozarkodina crispa* (Walliser, 1964) beta Morph.
   Walliser and Wang, 1989; Wang, pl. 1, figs 1–5
   (P$_1$ element).

Material. Several specimens of all elements.

Diagnosis. 'Sp [= P$_1$] element of *Ozarkodina crispa* with an asymmetrical, broadly expanded basal cavity. Oral margin of the blade straight over the whole length or higher at its anterior end, with or without middle furrow, ending at or anterior to the posterior margin of the cavity.' (Walliser and Wang 1989, p. 114, extended from original diagnosis given by Walliser, 1964, p. 75).

Remarks. This is a globally widespread index species for the uppermost Ludlow. The posterior edge of the blade on the P$_1$ element is characteristically erect, gently concave, or posteriorly inclined in lateral view. Different morphotypes of the P$_1$ element have been distinguished on the basis of material from south China (Walliser and Wang 1989) and from the northern East Baltic (Viira and Aldridge 1998). The beta morph of Walliser and Wang (1989) has a continuous furrow in the oral margin, but this does not show the pronounced sinuous widening of the furrow that is evident in the gamma morph. A suite of P$_2$ and ramiform elements occurs in association with the P$_1$ elements in the collections we have examined in south China (see Pl. 22, figs 6–8, 11–12, 15–19), and these elements are here tentatively assigned to the same apparatus. The P$_2$ element is long, with a nearly straight aboral edge; denticles are slender, crowded and all inclined posteriorly. The ramiform elements all have a narrow zone of inverted basal cavity along the processes and possess tall, widely spaced denticles with the development of occasional small denticles in the gaps; some specimens are present with genuine alternating denticulation (e.g. Pl. 22, fig. 18).

Occurrence. Sashuiyan Formation and lower part of Maliuqiao Formation, Erlangshan district, Sichuan.

Ozarkodina guizhouensis (Zhou, Zhai and Xian, 1981)
Plate 23, figures 1–14

*1981 *Spathognathodus guizhouensis* Zhou *et al.*, p. 137,
   pl. 2, figs 3–4 (P$_1$ element).
1981 *Neoprioniodus magnus* Zhou *et al.*, p. 135, pl. 1,
   figs 28–29 (M element).

?1981 *Ozarkodina duyunensis* Zhou, *et al.*, p. 136, pl. 2, figs 26–27 (P$_2$ element).

?1981 *Spathognathodus robustissimus* Zhou *et al.*, p. 138, pl. 2, figs 10–11 (P$_1$ element).

1983 *Spathognathodus guizhouensis* Zhou, Zhai et Xian, 1981; Zhou and Zhai, p. 296, pl. 68, fig. 5a–b (P$_1$ element).

?1983 *Ligonodina variabilis* Nicoll et Rexroad, 1968; Zhou and Zhai, p. 279, pl. 66, fig. 11. (S$_{3-4}$ element).

1983 *Neoprioniodus magnus* Zhou, Zhai et Xian, 1981; Zhou and Zhai, p. 283, pl. 66, fig. 19a–b (M element).

?1983 *Ozarkodina duyunensis* Zhou, Zhai et Xian, 1981; Zhou and Zhai, p. 288, pl. 67, fig. 11 (P$_2$ element).

1983 *Spathognathodus luoshuichongensis* Zhou and Zhai, p. 297, pl. 68, fig. 7 (P$_1$ element).

1989 *Ozarkodina guizhouensis* Zhou, Zhai et Xian, 1981; Yu *in* Jin *et al.*, p. 104, pl. 7, fig. 10a–b (P$_1$ element).

v 1996 *Ozarkodina guizhouensis* (Zhou *et al.*); Wang and Aldridge, pl. 3, fig. 1 (P$_1$ element).

v 2002 *Ozarkodina guizhouensis* (Zhou *et al.*); Aldridge and Wang, fig. 64A. [copy of Wang and Aldridge 1996, pl. 3, fig. 1] (P$_1$ element).

Diagnosis. *Ozarkodina* species in which all elements are robust with a very shallow basal cavity containing a slit-like groove; the cavity is partly inverted under the processes. P$_1$ element commonly arched, with denticles fused nearly to the tips; posterior process much shorter than anterior process. P$_2$, M and S elements with very prominent cusps; S$_{1-2}$ and S$_{3-4}$ elements twisted. S$_0$ with a short, denticulate posterior process.

Material. P$_1$, 87; P$_2$, 44; M, 38; S$_0$, 7; S$_{1-2}$, 9; S$_{3-4}$, 39; plus additional material from TT samples.

Description. P$_1$ element carminate. Blade robust, thick, gently arched in lateral view, straight or slightly bowed inwards in oral view; it bears a row of fused denticles with short, triangular free tips. Cusp posteriorly located, anterior process about twice length of posterior process. Cusp sometimes prominent, but may only be slightly larger than adjacent denticles, particularly in larger specimens. Anterior process with 6–11 denticles, low distally and commonly with two or three broader, higher denticles in the central portion. Posterior process with 3–5 denticles, regularly decreasing in height distally. Shallow basal cavity situated somewhat posterior of centre, forms a slit-like groove; usually lacking flared lips, although in some specimens very slightly flaring is visible. Groove extends anteriorly and posteriorly for a short distance; a narrow zone of inverted basal cavity occurs beneath the posterior process and, in larger specimens, beneath the anterior process. White matter difficult to discern, but appears to fill all denticle tips.

P$_2$ element angulate. Blade arched, curving inwards at posterior end or near the centre; thick and robust, bearing a row of short, fused, laterally compressed denticles. Cusp prominent

and posteriorly inclined. Anterior process longer than posterior process and bears seven to nine denticles which increase in size proximally. Posterior process short, inwardly curved, and bearing three or four posteriorly inclined denticles of subequal size. Basal cavity very shallow forming a slit-like groove; narrow inverted zone developed proximally on both processes.

M element makellate. Cusp large, stout, inclined and slightly twisted, extending downwards to form short triangular anticusp. Longer process of even height, directed at about 120° from the cusp, bearing a row of denticles of subequal size, commonly fused. Basal cavity shallow and broad beneath the cusp with a slit-like groove that extends beneath the longer process, but does not reach the distal tip.

S$_0$ element alate. Cusp tall, inclined posteriorly, with sharp anterior and posterior edges; transverse section biconvex. Posterior process short, laterally compressed, adenticulate or more commonly with one or two small denticles; height of process decreases rapidly distally. Lateral processes of even height, bearing subequal denticles which may be fused. Basal cavity very shallow beneath cusp with a small pit; becoming inverted along lower surfaces of lateral processes to form prominent keel.

S$_{1-2}$ element digyrate. Cusp erect or slightly inclined posteriorly, somewhat twisted with two sharp edges. Two processes of slightly unequal length arise from these edges, directed outwards and downwards and curving in towards each other. Shorter process with three to four denticles; proximal or second denticle conspicuously larger than the rest. Longer process with three to four discrete, more peg-like denticles. Basal margin of each process inverted to produce a keel.

S$_{3-4}$ element bipennate. Cusp large, posteriorly inclined and slightly twisted with a biconvex cross-section and sharp posterior and antero-lateral edges. Antero-lateral process directed strongly downwards and bowed inwards, bearing up to six discrete denticles. Posterior process slightly flexed and longer, bearing four to seven discrete or somewhat fused, confluent denticles; denticles commonly small proximally, with third and fourth denticles often conspicuously larger than the remainder. Basal margin of both processes inverted.

Remarks. *Spathognathodus luoshuichongensis* Zhou and Zhai (1983, p. 297, pl. 68, fig. 7) shows similar arching, even denticulation and inverted cavity to the P$_1$ element of *O. guizhouensis*, but is smaller; it probably represents a juvenile of the same species; *S. robustissimus* Zhou *et al.* (1981, p. 138, pl. 2, figs 10–11) is also possibly conspecific. The specimen illustrated as *Ozarkodina ziguiensis* Ni by Ni (1987, p. 424, pl. 62, fig. 4) is also similar, but has a less arched aboral margin and fewer denticles than typical specimens of *O. guizhouensis*. The specimen figured as *Ozarkodina duyuensis* Zhou *et al.* by Zhou *et al.* (1981, pl. 2, figs 26–27) and by Zhou and Zhai (1983, pl. 67, fig. 7) is robust with a clearly inverted basal margin and appears to be a specimen of the P$_2$ element of *O. guizhouensis*.

Spathognathodus luomianensis Zhou *et al.* (1981, p. 137, pl. 2, figs 16–17) is a robust P$_1$ element similar to that of

O. guizhouensis, but is straight and has a recessive basal cavity along most of the aboral margin; it appears to represent a distinct separate species. *Spathognathodus jigulingensis* Zhou and Zhai (1983, p. 296, pl. 68, fig. 10, a–b) is also straight with a long recessive basal margin and may be synonymous with *S. luomianensis*.

Occurrence. Lower member, Xiushan Formation, Leijiatun section, Shiqian County, Guizhou, and from many other localities in Guizhou (Zhou *et al.* l981; Zhou *et al.* 1985); Wangjiawan Formation, Yushitan section, Ningqiang, Shaanxi (sample TT 380; also Ding and Li 1985); Modaoya Formation (Liu *et al.* 1993) and Hanjiadian Formation (Zhou and Yu 1984), Sichuan.

Ozarkodina aff. *hassi* (Pollock, Rexroad and Nicoll, 1970)
Plate 23, figures 15–27

aff. 1970 *Spathognathodus hassi* Pollock *et al.*, p. 760, pl. 111, figs 8–12 (P_1 element).
aff. 1990 *Ozarkodina hassi* (Pollock, Rexroad and Nicoll, 1970); Armstrong, p. 92, pl. 13, figs 10–12, 14–16, *non* fig. 13 (multielement).
aff. 2002 *Ozarkodina hassi* (Pollock, Rexroad and Nicoll, 1970); Zhang and Barnes, p. 28, figs 13.31–13.37 (multielement).

Material. P_1, 13; P_2, 11; M, 3; S_0, 3; S_{1-2}, 3; S_{3-4}, 3.

Description. P_1 element carminate. Blade straight or slightly bowed. Anterior process longer than posterior, bearing four to seven denticles, which are fused in adult specimens but discrete in juveniles. Cusp prominent, posteriorly inclined. Posterior process short, with two to four partly fused denticles. All denticles laterally compressed. Basal cavity small, shallow, with a sharp tip beneath cusp.

P_2 element angulate. Blade short, arched and slightly bowed, bearing three or four laterally compressed denticles on each process. Cusp prominent; posterior process lower than anterior process. Cavity minute, with only slight flaring below cusp.

M element dolabrate. Cusp erect, with sharp anterior and posterior edges. Posterior process straight, strongly directed downwards, bearing discrete denticles that tend to be taller distally. Anterior process very short, adenticulate or with a few small, fused denticles. Basal cavity small, flared inwards beneath cusp, but narrowing rapidly below posterior process.

S_0 element alate. Cusp tall, inclined gently posteriorly, with sharp lateral edges. Lateral processes symmetrical, bearing three to four compressed, partially fused denticles; there is no posterior process. Posterior edge of small basal cavity slightly posteriorly flared beneath cusp, extending as a tapering, shallow groove to tips of lateral processes.

S_{1-2} element digyrate. Cusp erect, compressed, not strongly prominent on the one incomplete specimen examined. Both processes broken; shorter process with two preserved denticles, taller than the longer process which bears four discrete denticles. Basal cavity minute, very slightly flared beneath cusp.

S_{3-4} element bipennate; anterior process curved gently inwards and directed a little downwards to form 150° angle with posterior process, bearing five denticles that are crowded near the slender cusp but become discrete distally. Posterior process incomplete, with four preserved denticles, crowded proximally. Basal cavity minute beneath cusp.

Remarks. There are several differences in the Chinese material from specimens previously assigned to *O. hassi*. Denticles on the P_1 element are fewer in number and less crowded than on the holotype and associated specimens

---

EXPLANATION OF PLATE 23

Figs 1–7. *Ozarkodina guizhouensis* (Zhou *et al.* 1981). Lower member, Xiushan Formation, Leijiatun Section, Shiqian County, Guizhou, Sample Shiqian 15. 1, 149983, P_1 element, inner lateral view. 2, 149984, P_1 element, inner lateral view. 3, 149985, P_2 element, inner lateral view. 4, 149986, S_{1-2} element, posterior view. 5, 149987, M element, inner lateral? view. 6, 149988, S_0 element, lateral view. 7, 149989, S_{3-4} element, inner lateral view.

Figs 8–9, 11–14. *Ozarkodina guizhouensis* (Zhou *et al.* 1981). Lower member, Xiushan Formation, Leijiatun Section, Shiqian County, Guizhou, Sample TT 752b. 8, 149990, P_1 element, inner lateral view. 9, 149991, P_1 element, inner lateral view. 11, 149992, M element. inner lateral? view. 12, 149993, P_2 element, outer lateral view. 13, 149994, S_{3-4} element, inner lateral view. 14, 149995, S_0 element, posterior view.

Fig. 10. *Ozarkodina guizhouensis* (Zhou *et al.* 1981). Wangjiawan Formation, Yushitan-Shizuigou Section, Ningqiang County, Shaanxi, Sample TT 380. 117115, P_1 element, inner lateral view.

Figs 15–21. *Ozarkodina* aff. *hassi* (Pollock *et al.*, 1970). Kuanyinchiao Bed, Leijiatun Section, Shiqian County, Guizhou, Sample Shiqian-1. 15, 149996, P_1 element, lateral view. 16, 149997, P_1 element, lateral view. 17, 149998, P_1 element, lateral view. 18, 149999, P_2 element, lateral view. 19, 150000, P_2 element, lateral view. 20, 150001, S_{1-2} element, posterior view. 21, 15002, S_{1-2}? element, posterior view.

Figs 22–27. *Ozarkodina* aff. *hassi* (Pollock *et al.*, 1970). Kuanyinchiao Bed, Leijiatun Section, Shiqian County, Guizhou, Sample Shiqian-2. 22, 150003, P_1 element, lateral view. 23, 150004, P_1 element, lateral view. 24, 150005, M element, lateral? view. 25, 150006, P_2 element, lateral view. 26, 150007, S_0 element, posterior view. 27, 150008, S_{3-4} element, inner lateral view.

Figs 1–14, ×40; figs 15–27, ×140.

PLATE 23

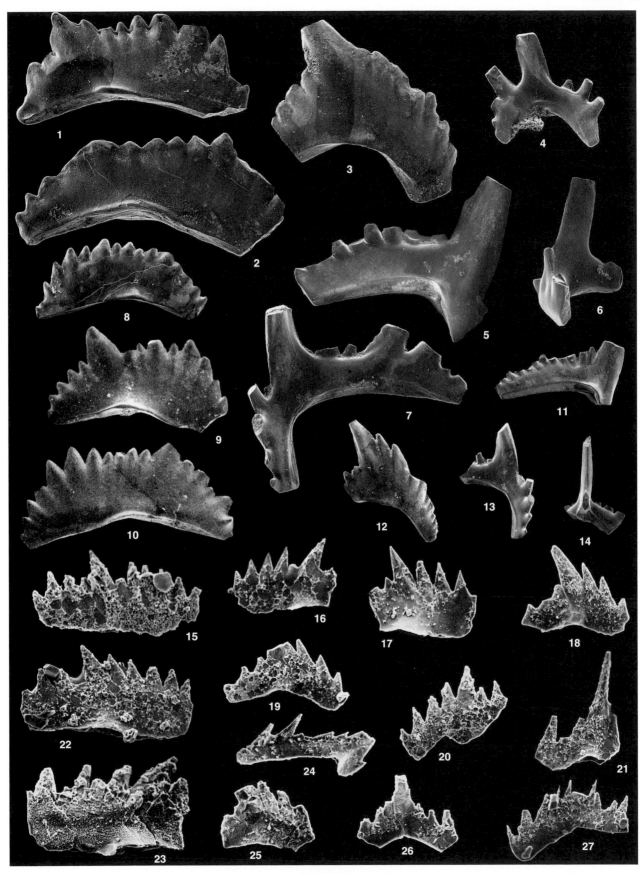

WANG and ALDRIDGE, *Ozarkodina*

(Pollock *et al.* 1970, pl. 111, figs 8–12). The P$_2$ element also has fewer, less crowded denticles than on the specimens referred to *O. edithae* by Pollock *et al.* (1970, pl. 113, figs 1–4), and the denticle tips on the anterior process do not form a straight line. The Chinese specimens are closer to those figured from Greenland by Armstrong (1990, pl. 13, figs 10–12). Given the differences from the type material, we have not assigned our specimens to *O. hassi*, but they are conceivably conspecific, and we do not have enough material to form the basis of a separate species.

The M element we identify is distinctly different from the specimen illustrated by Armstrong (1990, pl. 13, fig. 13), which is more planate with a posterior process. The basal cavity of Armstrong's specimen differs significantly from that of the other ramiform elements, being narrow and groove-like, and we consider that it comes from a different apparatus.

Occurrence. Kuanyinchiao Bed, Leijiatun section, Shiqian County, Guizhou.

Ozarkodina obesa (Zhou, Zhai and Xian, 1981)
Plate 24, figures 1–6

1981 *Spathognathodus guiyangensis* Zhou *et al.*, p. 137, pl. 2, figs 5–6 (P$_1$ element).
*1981 *Spathognathodus obesus* Zhou *et al.*, p. 137 pl. 2, figs 18–19 (P$_1$ element).
1983 *Paraspathognathodus guiyangensis* (Zhou, Zhai et Xian, 1981), Zhou and Zhai, p. 293, pl. 68, figs 11–12 (P$_1$ element).
1983 *Paraspathognathodus obesus* (Zhou, Zhai et Xian, 1981), Zhou and Zhai, p. 293, pl. 68, fig. 13a–b (P$_1$ element).
1987 *Paraspathognathodus guiyangensis* (Zhou, Zhai and Xian, 1981); An, p. 201, pl. 32, figs 9–10 (P$_1$ element).
2002 *Ozarkodina guiyangensis* Zhou, Zhai and Xian, 1981; Männik, p. 87, figs 7H–I, K–L, 8A–D, G, J. (P$_1$ and P$_2$ elements).

Diagnosis. P$_1$ element with a very large anterior denticle; cusp and other denticles of subequal size, with no gap between anterior and adjacent denticle. Denticle row subparallel to aboral edge, decreasing a little in height towards posterior end; cusp not prominent. Basal cavity long and shallow, rather slit-like, flaring most widely towards anterior end.

Material. P$_1$, 54; P$_2$, 11; M, 8; S$_{1-2}$, 6; S$_{3-4}$, 7.

Description. P$_1$ element carminate. Blade straight, thick, with a small cusp that is only slightly larger than adjacent denticles; in larger specimens blade thickened longitudinally at about midheight. Anterior denticle much larger than remainder,

broad-based and high, inclined or curved slightly posteriorly; anterior edge straight, adenticulate, normally with a small geniculation above the antero-aboral corner. All denticles partly fused, with free tips; no gap in denticulation between anterior denticle and its neighbour. Denticles number seven to at least ten, of subequal height apart from the anteriormost, commonly a little smaller and more crowded between cusp and anterior denticle. Denticles posterior to cusp as broad as or broader and higher than cusp, decreasing in height gradually posteriorly and becoming more posteriorly inclined. Basal cavity shallow and elongate, flaring most widely beneath the third denticle from anterior end, continuing as a narrowing groove to posterior tip. Aboral edge straight beneath anterior denticle, where it makes an angle of slightly greater than 90° with anterior edge; there is a sharp upward deflection of the profile beneath the widest extent of the cavity lips, and the profile is then straight to posterior tip.

P$_2$ element angulate. All specimens in our collections small and delicate, probably juvenile. Cusp prominent, tall and slender. Anterior and posterior processes almost all broken, but with confluent denticles fused almost to the tips. Basal cavity with only slightly flared lips beneath cusp.

M element dolabrate with erect cusp, short anticusp. Single process broken on all specimens, but probably long with erect denticles. Cavity relatively prominent below cusp, flaring and extending as prominent groove beneath process.

S$_{1-2}$ and S$_{3-4}$ elements in our collection small and delicate; all specimens broken, but with slender cusps, confluent denticles and minute basal cavities. S$_0$ element not recognized.

Remarks. Zhou *et al.* (1981) erected two species for P$_1$ elements with very prominent anterior denticles, *Spathognathodus guiyangensis* and *S. obesus*. Zhou and Zhai (1983) placed them both in their new genus *Paraspathognathodus*, with *S. obesus* as the type species. The primary differences between the two species are in the more robust nature of *S. obesus*, which also has fewer denticles and a less elongate cavity. These differences are relatively minor and show intergradation, and they may well be related simply to the size of the specimens; therefore, we consider the two to be synonymous. We have retained the name 'obesa', as this preserves the name for the type species of *Paraspathognathodus*. We currently, however, include the species within the genus *Ozarkodina*; if future phylogenetic analysis indicates that this species is a member of a distinct clade, then *Paraspathognathodus* would be an available name for this clade.

There are similarities to the P$_1$ element of *O. wangzhunia* sp. nov., which also has a prominent, although less broad and more erect, anterior denticle, but differs in the arched upper margin of the denticle row and the clear separation between the anterior denticle and its neighbour. Juvenile specimens of *O. obesa* may resemble *O. wangzhunia* more closely, especially in displaying an arching of the denticle row. There is also a clear relationship with the specimen figured as the Pa element of *Ozarkodina* aff. *broenlundi* by Männik (1983, fig. 5R), which has very similar denticulation to the P_1 element of *O. obesa*; the specimen from Severmaya Zemlya, however, has a more strongly posteriorly inclined anterior denticle and displays a strong lateral ledge below the denticle row, so it is currently not included in synonymy with *O. obesa*; Männik (2002) did not include this specimen in synonymy with the specimens from Severnaya Zemlya he identified as *O. guiyangensis*.

Männik (2002, fig. 7I, L) recognized probable P_2 elements; these are angulate, with a distinct, but not large, cusp and relatively long processes bearing slender, fused denticles. The specimens of the P_2 element in our collection are all broken juveniles, but appear to have shorter processes and a more prominent cusp.

Occurrence. Xiangshuyuan and Leijiatun formations, Leijiatun section, Shiqian County, Guizhou Huanggexi Formation, Huanggexi section, Daguan County, Yunnan. Also reported from the Shanggaozhaitian Group in Wudang, Guizhou (Zhou *et al.* 1981; Zhou and Zhai 1983) and the Hangjiadian Formation at Zunyi, Guizhou (An 1987).

Ozarkodina aff. *obesa* (Zhou, Zhai and Xian, 1981)
Plate 24, figures 7–8

Remarks. A single, large P_1 element from sample Ningqiang 7 shows a very prominent anterior denticle of the style of *O. obesa*, but has three small denticles down the anterior edge. The remainder of the blade bears ten denticles of varying size that form a gentle arch similar to that of specimens of *O. wangzhunia* sp. nov. There is an arched, longitudinal lateral ridge just below midheight of the blade and the cavity is slit-like, similar to that of *O. broenlundi*, and filled with a basal body. The distinctive characters of this specimen may partly be a consequence of its large size, and its relationships are not clear on the basis of this single individual.

Occurrence. Wangjiawan Formation, bed 1, Yushitan section, Ningqiang County, Shaanxi.

Ozarkodina parahassi (Zhou, Zhai and Xian, 1981)
Plate 24, figures 9–25

1981 *Spathognathodus parahassi* Zhou *et al.*, p. 139, pl. 2, figs 1–2 (P_1 element).
1981 *Spathognathodus wudangensis* Zhou *et al.*, p. 139, pl. 2, figs 14–15 (P_1 element).
1983 *Ozarkodina wudangensis* (Zhou, Zhai et Xian, 1981); Zhou and Zhai, p. 289, pl. 67, fig. 14a–b (P_1 element).

1983 *Spathognathodus parahassi* Zhou, Zhai et Xian, 1981; Zhou and Zhai, p. 298, pl. 68, fig. 9a–b (P_1 element).
?1985 *Spathognathodus hassi* Pollock, Rexroad and Nicoll, 1970; Qiu, p. 31, pl. 1, figs 21–22 (P_2, P_1 elements).
?1988 *Spathognathodus hassi* Pollock, Rexroad and Nicoll; Qiu, pl. 1, fig. 1 (P_1 element).
1989 *Ozarkodina hassi* (Pollock, Rexroad et Nicoll); Yu *in* Jin *et al.*, pl. 4, fig. 11 (P_1 element).
?1992 *Ozarkodina hassi* (Pollock, Rexroad and Nicoll); Qian *in* Jin *et al.*, pl. 2, fig. 9 (P_1 element).

Diagnosis. P_1 and P_2 elements each with a prominent cusp and relatively few denticles on anterior and posterior processes. Denticles at anterior end of P_1 element typically larger and higher than those close to cusp, but not forming a prominent crest; basal cavity flared beneath cusp, with lanceolate lips.

Material. P_1, 11; P_2, 2; ?M, 1; ?S_0, 2; ?S_{1-2}, 2; ?S_{3-4}, 1; plus additional material from TT samples.

Description. P_1 element carminate, with prominent, posteriorly inclined cusp. Anterior process with 4–6 denticles, distal one or two typically slightly broader and higher than proximal denticles. Posterior process with 3–6 erect or slightly posteriorly inclined denticles, with one or two in the central portion of the process commonly a little larger than the others. Basal margin straight with a lanceolate flare to the basal cavity with the widest point anterior to cusp.

P_2 element angulate. Cusp prominent, processes short. Anterior process with four or five denticles of subequal size, becoming higher proximally; posterior process lower, with up to five posteriorly inclined denticles; on the most complete specimen (Pl. 24, fig. 14) second and third denticles from cusp larger than remainder. Aboral edge straight to gently sigmoidally curved; cavity with flared lips beneath cusp.

Ramiform elements associated with the P_1 and P_2 specimens in the same samples show generally discrete denticles of subcircular transverse section (Pl. 24, figs 15–18, 22); the exception is the associated S_0 (Pl. 24, fig. 19), which has compressed denticles. However, other *Ozarkodina* species, including *O. obesa*, occur in the same samples, and assignment of ramiform elements is equivocal.

Remarks. There is variation in the relative lengths of anterior and posterior processes on the P_1 element in the populations we have examined, and these appear to encompass specimens of the type referred to *S. parahassi* and *S. wudangensis* by Zhou *et al.* (1981). The denticulation of the holotype of *O. wudangensis* is not completely clear from the illustration provided by Zhou *et al.* (1981, pl 2, figs 14–15), but the cavity shape is similar to our specimens, as is the relatively erect orientation of the posterior denticles.

The P_1 element is similar to that of *O. hassi* (Pollock *et al.* 1970, pl. 111, figs 8–12) in the prominent cusp and lanceolate basal cavity, but differs in having far fewer denticles on the anterior process. There are also close similarities to *O. parainclinata*, and the two might be synonymous, but *O. parainclinata* is distinguishable on the basis of the lower anterior denticles and the more extended, less laterally expanded lips to the basal cavity. There is also some similarity to *O. pirata*, and the two may occur in the same sample, but P_1 elements of *O. pirata* show a less prominent cusp and the denticle height decreases distally both anteriorly and posteriorly.

Occurrence. Leijiatun Formation, Leijiatun section, Shiqian County, Guizhou; Huanggexi Formation, Huanggexi section, Daguan County, Yunnan. Also reported from the Shanggaozhaitian Group, Wudang section, Guizhou (Zhou *et al.* 1981).

Ozarkodina cf. *parainclinata* (Zhou, Zhai and Xian, 1981)
Plate 25, figures 1–12

cf. 1981 *Spathognathodus parainclinatus* Zhou *et al.*, p. 138, pl. 2, figs 8–9 (P_1 element).
?1987 *Spathognathodus* cf. *parahassi* Zhou, Zhai and Xian; An, p. 203, pl. 32, fig. 20 (P_1 element).

Material. P_1, 6; P_2, 9; M, 9; S_0, 3; $S_{1–2}$, 4, $S_{3–4}$, 15.

Remarks. A set of generally large specimens occurs in association with specimens of *O. pirata*, and shares some characteristics, particularly in the general morphology of the P_1 element; however, these specimens do seem to represent a distinct apparatus. The P_1 element is elongate, with a straight aboral edge and with a narrow zone of inverted basal cavity along the posterior half of the aboral margin. The features generally accord with those described by Zhou *et al.* (1981, p. 138) for '*S.*' *parainclinata*: a blunt, round anterior end and pointed posterior end; cusp projecting but not strong and large, situated one-third of the distance from the posterior end; anterior blade high, posterior blade low. There are, differences, though, in the greater number of denticles in several specimens we have studied, and in the inverted basal cavity. The illustrations provided by Zhou *et al.* (1981, pl. 2, figs 8–9) do not show the cavity clearly, so we are not certain that it is different; for the present we simply compare our specimens with the holotype of *parainclinata*. There are also strong similarities between these specimens and *O. guizhouensis*, the primary difference being the concave basal edge of the P_1 element of the latter; it is possible that the specimens here referred to *O.* cf. *parainclinata* represent an early form of *O. guizhouensis*.

The other elements are distinctly different from their counterparts in the apparatus of *O. pirata*, and all show an area of inverted basal cavity along the aboral margin. The P_2 element is longer, more strongly arched and bears more denticles, typically 5–6 on the anterior process and 4–5 on the posterior process. The M element has an acutely angled aboral margin below the cusp, which is tall and twisted. Denticles on the S elements are more robust and may be a little more discrete.

EXPLANATION OF PLATE 24

Figs 1–2. *Ozarkodina obesa* (Zhou *et al.*, 1981). Upper Huanggexi Formation, Huanggexi Section, Daguan County, Yunnan, Sample TT 1136. 1, 150009, P_1 element, lateral view. 2, 150010, P_1 element, lateral view.

Figs 3–4. *Ozarkodina obesa* (Zhou *et al.*, 1981). Uppermost Huanggexi Formation, Huanggexi Section, Daguan County, Yunnan, Sample TT 1141b. 3, 150011, P_1 element, lateral view. 4, 150012, P_1 element, lateral view.

Figs 5–6. *Ozarkodina obesa* (Zhou *et al.*, 1981). Xiangshuyuan Formation, Leijiatun Section, Shiqian County, Guizhou, Sample Shiqian 4. 5, 150013, P_1 element, lateral view. 6, 150014, M? element, inner lateral? view.

Figs 7–8. *Ozarkodina* aff. *obesa* (Zhou *et al.*, 1981). Wangjiawan Formation, Yushitan Section, Ningqiang County, Shaanxi, Sample Ningqiang 7. 150015, P_1 element, lateral view, and close-up to show striae on denticles and wear facet developed on left-hand denticle.

Figs 9–19. *Ozarkodina parahassi* (Zhou *et al.*, 1981). Uppermost Huanggexi Formation, Huanggexi Section, Daguan County, Yunnan, Sample TT 1141b. 9, 150016, P_1 element, lateral view. 10, 150017, P_1 element, lateral view. 11, 150018, P_1 element, lateral view. 12. 150019, P_1 element, lateral view. 13, 150020, P_1 element, lateral view. 14, 150021, P_2 element, lateral view. 15, 150022, ?M element, inner lateral? view. 16, 150023, ?M element, inner lateral? view. 17, 150024, ?$S_{3–4}$ element, inner lateral view. 18, 150025, ?$S_{1–2}$ element, posterior view. 19, 150026, ? S_0 element, posterior view.

Figs 20–25. *Ozarkodina parahassi* (Zhou *et al.*, 1981). Upper Huanggexi Formation, Huanggexi Section, Daguan County, Yunnan, Sample TT 1140. 20, 150027, P_1 element, lateral view. 21, 150028, P_2 element, lateral view. 22, 150029, ?$S_{1–2}$ element, posterior view. 23, 150030, P_1 element, lateral view. 24, 150031, P_1 element, lateral view. 25, 150032, P_1 element, oblique lateral view.

All figures ×80, except fig. 8, ×220.

PLATE 24

WANG and ALDRIDGE, *Ozarkodina*

Two of the specimens referred by Yu (*in* Jin *et al.* 1989, pl. 7, figs 7, 14a–b only) to *Ozarkodina yanheensis* Zhou and Zhai, 1983, are very similar to the specimens we include in *O*. cf. *parainclinata*. The specimen figured as *S*. cf. *parahassi* by An (1987, pl. 32, fig. 20) is broken, but does not display the prominent cusp of the P_1 element of *O. parahassi*; the denticulation is similar to the specimens we refer here to *O*. cf. *parainclinata*, but the cavity appears to be rather more flared.

Occurrence. Leijiatun Formation, Leijiatun section, Shiqian County, Guizhou (samples Shiqian 8, 9).

Ozarkodina paraplanussima (Ding and Li, 1985)
Plate 25, figures 13–27.

?1983 *Synprioniodina typica* Scholaub, 1971 [*sic*]; Zhou and Zhai, p. 299, pl. 68, fig. 16a–b (S_{1-2} element).
*1985 *Spathognathodus paraplanussimus* Ding and Li; p. 17, pl. 1, fig. 30 (P_1 element).
v 1996 *Ozarkodina planussima* (Zhou *et al.*); Wang and Aldridge, pl. 4, fig. 8 (P_1 element).
v 2002 *Ozarkodina planussima* (Zhou *et al.*); Aldridge and Wang, fig. 65H (copy of Wang and Aldridge 1996, pl. 4, fig. 8]) (P_1 element).

Diagnosis. (Modified after Ding and Li 1985, p. 17). P_1 element characteristic: thin with flat lateral surface and straight to slightly convex aboral edge; lower part of lateral surface with a concave trough parallel to aboral edge, basal cavity central, very slightly inflated; oral surface with 10–15 denticles, displaying prominent striae.

Material. P_1, 30; P_2, 6; M, 5, ?S_0, 4; ?S_{1-2}, 9; plus additional material from TT samples.

Description. P_1 element carminate. Blade straight or slightly bowed inwardly at posterior end, laterally compressed, with a straight to slightly sinuous aboral edge in lateral view. Blade bears a row of 10–15 fused, laterally compressed, sharp denticles, with cusp the same size as adjacent denticles or a little larger. Denticles of anterior process erect distally, becoming posteriorly inclined proximally, increasing in height towards cusp. Denticles of posterior process fused and inclined posteriorly, decreasing regularly in height distally. Blade beneath denticle row a little laterally expanded with a prominent ledge commonly developed parallel to basal margin. Basal cavity weakly expanded below cusp, extending as a shallow groove to tips of blade. All denticles bear fine but prominent striae parallel to denticle axis (Pl. 25, fig. 14).

P_2 element angulate. All specimens broken, but are thin, laterally compressed with an almost straight aboral edge and a very slight inflation to basal cavity beneath cusp. Cusp prominent, denticles crowded.

M element bipennate. Cusp prominent; processes form an angle of about 90°. Shorter process with a large denticle adjacent to cusp and two to five shorter denticles distally. Longer process straight or very gently curved, broken, with crowded or closely spaced slender, erect denticles. Basal cavity flared slightly beneath cusp, narrowing gradually beneath longer process. Cusp and all denticles with prominent striae.

?S_0 element. There are two candidates for the S_0 position of this apparatus. The most likely is represented by three specimens, two from sample Xuanhe 1 (Pl. 25, fig. 18) and one from Shiqian 17 (Pl. 25, fig. 26). All are broken, but are alate with a broad cusp proximally and crowded slender denticles on lateral processes. There is no posterior process, but there is a slightly posteriorly extended thickening immediately above the very

---

EXPLANATION OF PLATE 25

Figs 1–12. *Ozarkodina* cf. *parainclinata* (Zhou *et al.*, 1981). Leijiatun Formation, Leijiatun Section, Shiqian County, Guizhou, Sample Shiqian 9. 1, 150033, P_1 element, lateral view. 2, 150034, P_2 element, lateral view. 3, 150035, P_2 element, lateral view. 4, 150036, P_1 element, lateral view. 5, 150037, P_1 element, lateral view. 6, 150038, P_2 element, lateral view. 7, 150039, M element, inner lateral? view. 8, 150040, S_0 element, posterior view. 9, 150041, M element, inner lateral? view. 10, 150042, S_{1-2} element, posterior view. 11, 150043, S_{3-4} element, inner lateral view. 12, 150044, S_{3-4} element, inner lateral view.

Figs 13–14, 24–27. *Ozarkodina paraplanussima* (Ding and Li, 1985). Upper member, Xiushan Formation, Leijiatun Section, Shiqian County, Guizhou, Sample Shiqian 17. 13–14, NHM X1142, P_1 element, lateral view and close-up to show well-developed striae. 24, 150045, M element, inner lateral? view. 25, 150046, P_1 element, lateral view. 26, 150047, ?S_0 element, posterior view. 27, 150048, ?S_0 element, lateral view.

Figs 15–22. *Ozarkodina paraplanussima* (Ding and Li, 1985). Shenxuanyi Member, Xuanhe Section, Guangyuan County, Sichuan, Sample Xuanhe 1. 15, 150049, P_1 element, lateral view. 16, 150050, P_1 element, lateral view. 17, 150051, P_1 element, lateral view. 18, 150052, ?S_0 element, posterior view. 19, 150053, ?S_{1-2} element, posterior view. 20, 150054, M element, inner lateral? view. 21, 150055. P_2 element, lateral view. 22, 150056, ?S_0 element, posterior view.

Fig. 23. *Ozarkodina paraplanussima* (Ding and Li, 1985). Upper member, Xiushan Formation, Leijiatun Section, Shiqian County, Guizhou, Sample TT 740. 150057, P_1 element, lateral view.

Figs 1–12, 24–27, ×60; figs 13, 15–22, ×100; fig. 14, ×180.

PLATE 25

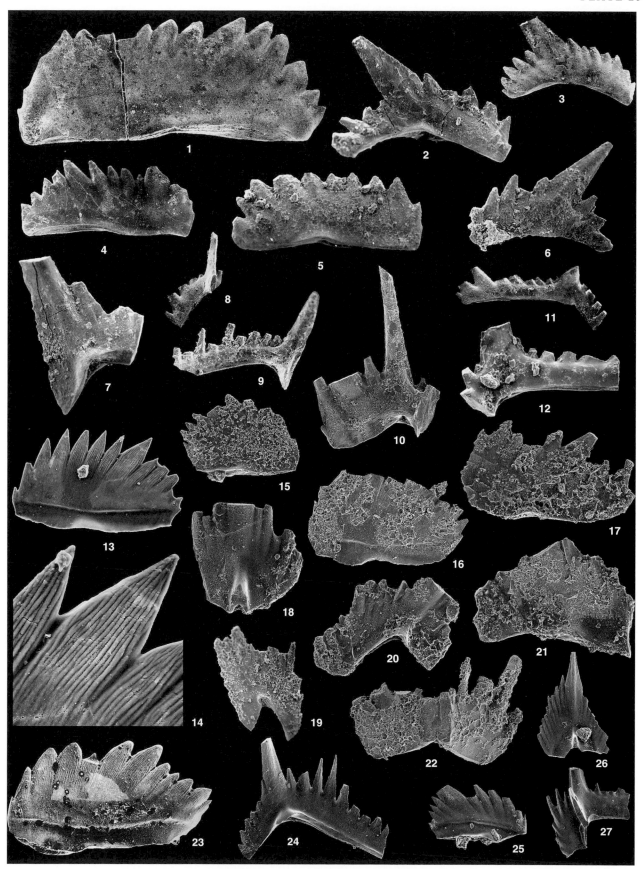

WANG and ALDRIDGE, *Ozarkodina*

small basal cavity. Cusp and all denticles with striae. A single specimen from sample Shiqian 17 (Pl. 25, fig. 27) is different, alate with erect cusp, well-developed denticulate posterior process and antero-laterally directed denticulate lateral processes; cusp and lateral denticles with surface striae; basal groove extends along all processes, with inverted basal margin, most extensive proximally.

?S_{1-2} element. Two equivocally assigned specimens in sample Xuanhe 1. Most likely (Pl. 25, fig. 19) digyrate, with two processes forming a very acute arch. Cusp short, broken. Shorter process with crowded, fused denticles; two proximal denticles relatively tall, distal denticles small and set successively downwards to produce aborally directed process. Longer process curved downwards with crowded denticles, tall proximally and decreasing steadily in size distally. Basal cavity very small, extended into a pinched lip posteriorly. Second specimen (Pl. 25, fig. 22), almost symmetrical, with lateral processes forming arch of 130°. Both processes with denticles that increase in size and inclination distally. Minute cavity beneath the broken cusp, extending as very narrow inverted zones beneath the processes.

S_{3-4} element not recognized.

Remarks. The P_1 element is very similar to that of *O. planussima* (*S. planussimus* of Zhou *et al.* 1981, p. 138, pl. 2, fig. 7), but that species is recorded as possessing 18–20 denticles, whereas those in the collections we have studied typically bear only 10–13 denticles. The holotype and only complete specimen originally described of *S. paraplanussimus* has 15 denticles and was differentiated from *S. planissimus* by Ding and Li (1985, p. 18) on the basis of denticle number and a nonlinear aboral edge. It is possible that these two species should be regarded as synonymous, especially as the denticulation on the illustration provided by Zhou *et al.* (1981) is unclear. However, the specimens we have studied accord more closely with the description of *O. paraplanussima*, so we assign them to that species here.

Ramiform elements associated with the P_1 elements are few in number and differ in morphology in differ-

ent samples, so the apparatus composition is uncertain. In particular, associated S_0 elements include specimens with and without a posterior process; only one of these, and possibly neither, can be from the *O. paraplanussima* apparatus. The associated M elements in different samples also show different denticulation, which may indicate intraspecific variation or may be the result of mis-assignment.

Occurrence. Upper member, Xiushan Formation, Leijiatun section, Shiqian County, Guizhou; Shenxuanyi Member, Ningqiang Formation, Yushitan section, Ningqiang, Shaanxi; lower part of Shenxuanyi Member, Xuanhe section, Guangyuan, Sichuan. Also reported from the Luomian Formation, Kaili County, Guizhou (Zhou *et al.* 1981).

Ozarkodina pirata Uyeno *in* Uyeno and Barnes, 1983
Plate 26, figures 1–12

> ?1970 *Ozarkodina* n. sp. B. Pollock *et al.*, p. 757, pl. 113, figs 9–11 (P_1 element).
> v 1979 *Ozarkodina* aff. *O. polinclinata* (Nicoll and Rexroad, 1969); Aldridge, p. 17, pl. 2, figs 1–5 (multielement).
>   1981 *Ozarkodina* n. sp. A Fåhraeus and Barnes, pl. 1, fig. 9 (P_1 element).
>   1981 *Ozarkodina* n. sp. C Uyeno and Barnes, pl. 1, fig. 2 (P_1 element).
> * 1983 *Ozarkodina pirata* Uyeno *in* Uyeno and Barnes, p. 21, pl. 1, figs 16–17, 21–25; pl. 2, figs 12–13, 19–24, 26–28, *non* fig. 25 (multielement).
> v 1990 *Ozarkodina pirata* Uyeno and Barnes, 1983; Armstrong, p. 94, pl. 13, figs 20–23; pl. 14, figs 1–5, 7–8, *non* fig. 6 (multielement).
>   2002 *Ozarkodina pirata* Uyeno, 1983; Zhang and Barnes, p. 29, figure 12.1–12.17 (multielement).

Diagnosis. See Zhang and Barnes (2002, p. 29), emended after Uyeno and Barnes (1983, p. 21).

EXPLANATION OF PLATE 26

Figs 1–12. *Ozarkodina pirata* Uyeno *in* Uyeno and Barnes, 1983. Leijiatun Formation, Leijiatun Section, Shiqian County, Guizhou, Sample Shiqian 9. 1, 150058, P_1 element, lateral view. 2, 150059, P_1 element, lateral view. 3, 150060, P_1 element, lateral view. 4, 150061, P_1 element, lateral view. 5, 150062, P_2 element, lateral view. 6, 150063, P_2 element, lateral view. 7, 150064, P_2 element, lateral view. 8, 150065, M element, inner lateral? view. 9, 150066, S_0 element, posterior view. 10, 150067, S_{1-2} element, posterior view. 11, 150068, S_{1-2} element, posterior view. 12, 150069, S_{3-4} element, inner lateral view.

Fig. 13. *Ozarkodina* cf. *planussima* (Zhou *et al.*, 1981). Upper member, Xiushan Formation, Leijiatun Section, Shiqian County, Guizhou, Sample Shiqian 20. 150070, P_1 element, lateral view.

Figs 14–23. *Ozarkodina wangzhunia* sp. nov. Leijiatun Formation, Leijiatun Section, Shiqian County, Guizhou, Sample Shiqian 8. 14, 150071, P_1 element, lateral view. 15–16, 150072, M element, inner lateral? and oblique aboral views. 17–18, 150073, P_2 element, lateral and oblique aboral views. 19–20, 150074, P_1 element, lateral and oblique aboral views (holotype). 21, 150075, S_{1-2} element, posterior view. 22, 150076, S_0 element, posterior view. 23, 150077, S_{3-4} element, inner lateral view.

All figures ×100 except fig. 13, ×80.

PLATE 26

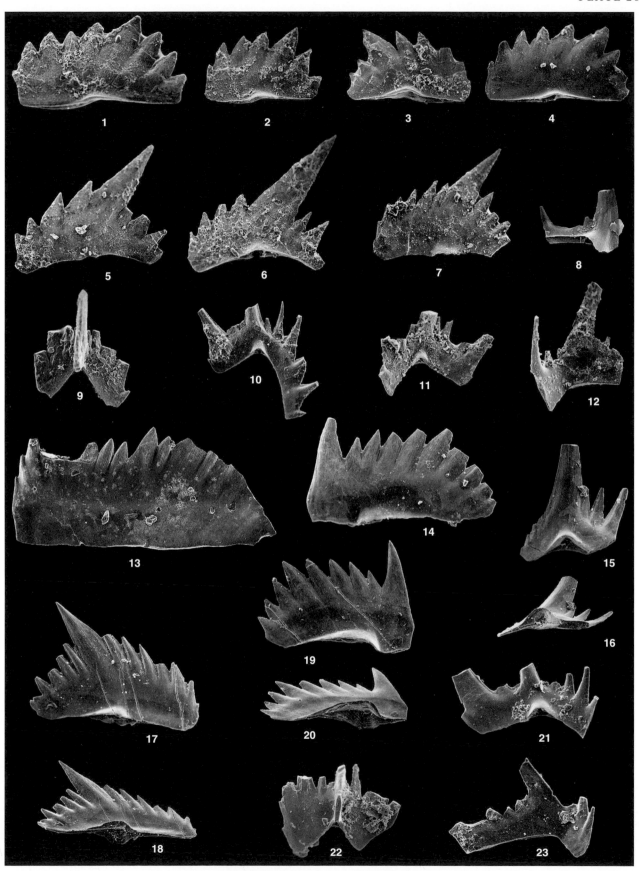

WANG and ALDRIDGE, *Ozarkodina*

Material. P_1, 70; P_2, 23; M, 5; S_0, 6; S_{1-2}, 6; S_{3-4}, 15.

Remarks. The P_1 elements from China are very closely similar to the type suite from Anticosti Island, Quebec (Uyeno and Barnes 1983, pl. 1, fig. 24, pl. 2, figs 12–13), but the P_2 elements have a slightly more prominent cusp. The remaining elements closely resemble those from the type locality, except that the arch between the lateral processes of the S_0 element is more acute in the Chinese material, as is the arch on the S_{1-2} element; the denticulation on the S_0 element is also more crowded.

The M element of *O. pirata* and that of *Pseudolonchodina fluegeli* are effectively indistinguishable, and some specimens may have been wrongly identified in our tables.

Zhang and Barnes (2002) included several specimens figured by Pollock *et al.* (1970, pl. 113, figs 5–8, 15–20, 22–24, pl. 114, figs 13–14) in synonymy with *O. pirata*. We have reservations about the inclusion of most of these, especially the P_1 elements (*Ozarkodina* n. sp. A Pollock *et al.* 1970, pl. 113, figs 5–8), which have more widely flared cavities and a more angular aboral edge than is typical for *O. pirata*. Zhang and Barnes (2002, p. 30) noted that their collections show different ontogenetic stages and that these encompass the morphologies of typical *O. pirata* and of *O.* n. sp. A, but their illustrations do not show any specimens comparable to the latter. In fact, the specimens referred to *Ozarkodina* n. sp. B by Pollock *et al.* (1970) more closely resemble the P_1 of *O. pirata*, and we include them equivocally in synonymy here. Some of the other elements they figure may also belong, but resolution of this probably requires reexamination of the Pollock *et al.* collections.

Occurrence. Leijiatun Formation, Leijiatun section, Shiqian County, Guizhou.

Ozarkodina cf. *planussima* (Zhou, Zhai and Xian, 1981)
Plate 26, figure 13

cf. 1981 *Spathognathodus planussimus* Zhou *et al.*, p. 138, pl. 2, fig. 7 (P_1 element).

Material. P_1, 2.

Remarks. Zhou *et al.* (1981, p. 138) described specimens with thin, flat blades, a linear aboral edge, a characteristic concave trough parallel to the basal margin and 18–20 denticles as *S. planussimus*. Ding and Li (1985, p. 17) subsequently recognized specimens that are similar in several characteristics, but have a nonlinear basal edge and fewer denticles (15 on the only complete specimen); they distinguished these as a new species, *S. paraplanussimus*. Most specimens in the collections we have studied are closer to

paraplanussimus (see above), but single broken P_1 elements from samples Shiqian 20 and Ningqiang 3 show similarities to *planussimus*, bearing at least 18 denticles and having a straight aboral edge. However, these differ from the holotype of *O. planussima* in being longer and in their relatively subtly developed concave trough above the aboral edge; given these differences and their incompleteness, we assign them here to *O.* cf. *planussima*.

Occurrence. Upper member, Xiushan Formation, Leijiatun section, Shiqian County, Guizhou; Yangpowan Member, Ningqiang Formation, Yushitan section, Ningqiang, Shaanxi.

Ozarkodina wangzhunia sp. nov.
Plate 26, figures 14–23

Derivation of name. After Wang Zhu, daughter of Wang Chengyuan.

Holotype. Specimen NIGPAS 150074 (Pl. 26, figs 19–20); P_1 element.

Type locality and horizon. Leijiatun section, Shiqian County, Guizhou Province; Leijiatun Formation, sample Shiqian 8.

Diagnosis. P_1 element with a prominent erect denticle at the anterior end of the blade, separated by a broad V-shaped space from the other denticles, the tips of which form an arch that decreases in height sharply posteriorly; basal cavity lanceolate with widest flaring close to anterior end. M element with a well-developed anticusp, S_0 element alate with short, adenticulate posterior process.

Material. P_1, 19; P_2, 29; M, 6; S_0, 1; S_{1-2}, 5; S_{3-4}, 10.

Description. P_1 element carminate. Blade straight, with a slightly undulose aboral edge. Cusp not prominent, of similar size to, or only very slightly larger than adjacent denticles, posteriorly inclined. Anterior end of blade marked by a single tall, slender, erect denticle that is much higher than the remainder. Anterior edge of blade inclined posteriorly for most of height, but with a small geniculation near the base, below which it is inclined anteriorly. Remainder of denticles posteriorly inclined, showing a gradual decrease in size away from cusp in both directions, the tips forming an arch that is more pronounced posteriorly. Denticles fused at their bases but discrete at tips. Lips of basal cavity lanceolate, symmetrical, occupying central portion of blade, but flaring most widely at their anterior end. Basal cavity shallow; in some specimens, a narrow groove extends posteriorly along aboral edge.

P_2 element angulate. Blade straight or slightly bowed inwards, with slender denticles fused nearly to their apices and a prominent cusp posterior to midlength. Anterior process with denticles of subequal height, posterior process becoming rapidly lower distally. Basal cavity with heart-shaped or lanceolate lips, widest beneath cusp, tapering anteriorly and posteriorly as a very narrow groove.

M element dolabrate. Cusp tall, bowed with sharp edges. Anticusp short but well-defined, adenticulate or with a few rudimentary denticles; longer process broken, but with at least two or three tall, discrete denticles, the most proximal of which may be fused to cusp. Basal cavity flared inwards, widest and deepest beneath cusp.

S_0 element alate. Cusp broken, but apparently erect with subtriangular cross-section. Lateral processes of subequal length enclosing an acute aboral arch, bearing three or four erect, tall denticles. Posterior process short and adenticulate. Basal cavity small and confined below cusp.

S_{1-2} element digyrate. Cusp inclined with subcircular or biconvex cross-section. Shorter process directed slightly downwards with two or three small denticles proximally and a single much larger denticle near distal tip. Longer process slightly curved, directed more strongly downwards and bearing about six denticles with the larger ones at midlength. Basal cavity slightly flared beneath cusp.

S_{3-4} element bipennate. Cusp tall, inclined posteriorly, with sharp anterior and posterior edges and with inner and outer lateral costae giving a nearly rhomboidal cross-section. Anterior processes short, bowed inwards and also directed downwards, with three to five discrete, compressed denticles. Posterior process long, with at least five discrete denticles. Aboral surface very narrow; no groove can be seen. Basal cavity very small beneath cusp.

Remarks. The P_1 element is similar to that of *O. obesa*, but differs in the arched upper margin of the denticle row, the less prominent, erect cusp and the flared lanceolate basal cavity. There is also a similarity to the specimen figured as the Pa element of *Ozarkodina* aff. *broenlundi* by Männik (1983, fig. 5R), but the latter has a posteriorly inclined anterior denticle that is fused to its neighbour and displays a strong lateral ledge below the denticle row and parallel to the aboral margin.

Occurrence. Leijiatun Formation, Leijiatun section, Shiqian County, Guizhou.

Ozarkodina waugoolaensis Bischoff, 1986
Plate 27, figures 1–12

?1983 *Ozarkodina yanheensis* Zhou and Zhai, p. 289, pl. 67, fig.15 (P_2 element).
1986 *Ozarkodina waugoolaensis* Bischoff, p. 145, pl. 23, figs 22–24, 26?, 27?, 28–45, pl. 24, figs 1–4, 5?, 6–10 (multielement).
1989 *Ozarkodina wudangensis* (Zhou, Zhai et Xian); Yu *in* Jin *et al.*, pl. 4, figs 4, 12, 13a–b, 17; pl. 5, fig. 9 (P_1 element).
p 1999 *Ozarkodina* aff. *broenlundi* Aldridge, 1979; Melnikov, p. 75, pl. 23, figs 12–16, *non* fig. 17 (P_1, P_2 and S_{1-2} elements).
p 1999 *Ozarkodina* ex. gr. *hassi* (Pollock, Rexroad et Nicoll); Melnikov, pl. 23, figs 20, 23–24, *non* figs 21–22 (P_1 element).
2002 *Ozarkodina waugoolaensis* Bischoff, 1986; Männik, p. 89, figs 9D–M, 10A–N, 11R–S, V (multielement).

Diagnosis. 'Pa (= P_1) element carminate with well-developed main and secondary cusps that display a different degree of posterior inclination and enclose a V-shaped gap in the upper outline between them; lower edge very gently sigmoidal or concave. Basal cavity lanceolate.' (Bischoff 1986, p. 145).

Material. P_1, 79; P_2, 58; M, 27; S_0, 4; S_{1-2}, 26; S_{3-4}, 33; plus additional material from TT samples.

Description. P_1 element carminate. Blade straight, elongate with aboral edge straight or nearly straight. Cusp broad and low, posteriorly inclined, situated one-third of length from posterior tip. Anterior blade higher than posterior, bearing five to eight broad, erect denticles; denticle anteriorly adjacent to cusp is of similar size to cusp and separated from it by a V-shaped space. Posterior blade with up to six denticles, inclined posteriorly, generally less prominent than those of the anterior, but relatively larger on smaller specimens (Pl. 27, fig. 4). Basal cavity shallow, with lanceolate lips, widest below large denticle adjacent to cusp and extending under cusp. Faint striations apparent on cusp and all denticles, sometimes worn away at tips.

P_2 element angulate. Cusp very prominent; anterior blade higher than posterior blade. Anterior blade with five or six broad, tall denticles, the largest adjacent to cusp. Posterior blade shorter, lower, with about six denticles, less prominent than on anterior blade. Basal edge straight; basal cavity shallow with lanceolate lips, flaring most widely below denticle anterior to cusp. Clear striations on cusp and all denticles, sometimes worn away at tips.

M element dolabrate. Cusp prominent, with a gently convex outer face and a more strongly convex inner face in transverse section. Longer process gently curved proximally, becoming straighter distally, broken on all specimens; bearing equant, closely spaced denticles, at least seven in number. Shorter process very short, adenticulate or with up to two tiny fused denticles. Basal margin of cusp extended into a prominent narrow lip on inner side. Basal cavity broad and shallow under cusp and longer process, developing a narrow zone of recessive margin along the process. Striations well developed on cusp and all preserved denticles of longer process.

S_0 element alate; cusp and posterior process broken on all specimens. Cusp apparently erect, subcircular in transverse section. Lateral processes deep, forming an arch of 90–100°, bearing compressed, fused denticles that display clear striations. Basal cavity a small pit beneath cusp, extending as a clear, broad zone of recessive basal margin, tapering distally, along aboral edge of both processes.

S_{1-2} element digyrate. Specimens all badly broken; asymmetrical with prominent cusp and shorter process that is higher and less strongly directed downwards than longer process. Both processes with fused, compressed denticles; denticles and cusp show striations. Basal cavity a small pit beneath cusp.

S_{3-4} element bipennate with a prominent posteriorly inclined cusp, which is laterally compressed with sharp anterior and posterior edges. Posterior process long and straight, bearing fused, compressed denticles that increase in size and in posterior inclination distally. Anterior process curving strongly downwards and a little inwards, bearing up to six compressed denticles. Basal cavity tiny beneath cusp. Faint striations apparent on cusp; denticles appear smooth, but only basal portion is preserved.

Remarks. The holotype of *O. waugoolaensis*, as designated by Bischoff (1986, p. 145, pl. 23, fig. 22), is a Pb element, not a Pa element as stated. It seems likely that there is an error of numbering on the plate and that it is the P_1 element illustrated in plate 23, figure 23 that was intended to be the holotype. The P_1 element is more elongate than that of *O. parahassi*, which also lacks the large denticle anteriorly adjacent to the cusp. The P_1 element of *O. broenlundi* is very similar in denticulation to that of *O. waugoolaensis*, and the two are probably closely related; specimens of *O. broenlundi* differ in the more slit-like basal cavity and in the high anterior denticles, although a single larger denticle is present near the anterior termination of some specimens of *O. waugoolaensis*. Specimens of the M element in the Chinese material have a distinct, downwardly directed lip extending from the axial line of the inner face of the cusp (Pl. 27, figs 6–7); this feature is not apparent in the specimens illustrated by Bischoff (1986, pl. 23, fig. 24 which shows an outer view, pl. 24, fig. 7), nor in the specimens illustrated by Männik (2002, fig. 10F–G), which differ from both the Chinese and Australian specimens in the slender denticles on the longer process and in the presence of two or three denticles on the shorter process. Some of the specimens assigned to the Sa and Sb positions of *O. waugoolaensis* by Bischoff (1986, pl. 23, figs 26–27, pl. 24, fig. 5) have discrete, peg-like denticles, quite different from the S_{1-2} and S_{3-4} elements recognized here, and they probably do not belong

to the apparatus. Although broken, the S_0 elements in association with the P_1 elements of *O. waugoolaensis* in the Chinese collection have a clear posterior process and a zone of recessive basal cavity. The ramiform elements of *O. broenlundi* Aldridge (1979, pl. 1, figs 22–25) have similar confluent denticulation to the ramiforms of *O. waugoolaensis* identified in the Chinese collections, but the S_0 element does not possess a posterior process.

The denticulation of the P_2 specimen named *Ozarkodina yanheensis* by Zhou and Zhai (1983) is similar to that of specimens of the P_2 element of *O. waugoolaensis* illustrated by Bischoff (1986). If they are conspecific, *yanheensis* would have priority as the specific name, but synonymy cannot be confirmed in the lack of information on P_1 elements associated with Zhou and Zhai's specimen.

Occurrence. Upper member, Xiushan Formation, Leijiatun section, Shiqian County, Guizhou; Yangpowan Member, Ningqiang Formation, Yushitan section, Ningqiang, Shaanxi; Shenxuanyi Member, Xuanhe section, Guangyuan, Sichuan.

Genus WURMIELLA Murphy, Valenzuela-Riós and Carls, 2004

Type species. *Ozarkodina excavata tuma* Murphy and Matti, 1983, pl. 1, figs 3–9 (= *Ozarkodina tuma* see Murphy *et al.* 2004, p. 8).

Diagnosis. Elements characterized primitively by processes without strong size variation of adjacent denticles. P_1 element with relatively small, narrow basal lobes that lack ornamentation. P_2 element with cusp much larger than other denticles and with slight asymmetry of the basal lobes so that the lobe on the inner side is elevated (modified from Murphy *et al.* 2004, p. 8).

Remarks. Murphy *et al.* (2004) introduced the genus *Wurmiella* to accommodate the apparatus widely referred to *Ozarkodina excavata* (Branson and Mehl, 1933a), and related taxa. Preliminary cladistic analyses (Donoghue *et al.*

EXPLANATION OF PLATE 27

Figs 1–12. *Ozarkodina waugoolaensis* Bischoff, 1986. Wangjiawan Formation, Yushitan Section, Ningqiang County, Shaanxi, Sample Ningqiang 7. 1, 150078, P_1 element, lateral view. 2, 150079, P_1 element, lateral view. 3, 150080, P_1 element, lateral view. 4, 150081, P_1 element, lateral view. 5, 150082, P_2 element, lateral view. 6, 150083, M element, inner lateral? view. 7, 150084, M element, inner lateral? view. 8, 150085, S_0 element, posterior view. 9, 150086, S_0 element, posterior view. 10, 150087, P_2 element, lateral view. 11, 150088, S_{1-2} element, posterior view. 12, 150089, S_{3-4} element, inner lateral view.

Figs 13–16. *Aulacognathus bullatus* (Nicoll and Rexroad, 1969). Hanjiadian/Jiujialu formation boundary, Zhangjiawan, Zheng'an County, Guizhou, Sample Zh1. 13, 150090, P_1 element, oral view. 14, 150091, P_1 element, oral view. 15, 150092, P_1 element, oral view. 16, 150093, P_1 element, oral view.

Figs 17–18. *Aulacognathus?* sp. Daluzhai Formation, Huanggexi Section, Daguan County, Yunnan, Samples TT 1169, TT 1165. 150094, 150095, P_1 elements, oral view.

All figures ×80.

PLATE 27

WANG and ALDRIDGE, *Ozarkodina, Aulacognathus*

2008) support the separation of *excavata* from *Ozarkodina*. Of the other taxa Murphy *et al.* (2004) assigned to *Wurmiella*, *O. hassi* does not appear to be phylogenetically associated (Donoghue *et al.* 2008) so is retained provisionally in *Ozarkodina*.

In the cladistic studies by Donoghue *et al.* (2008), *O. excavata* commonly forms a clade with *Yaoxianognathus*, and the possibility that the two are congeneric should be considered. *Yaoxianognathus*, erected by An (*in* An *et al.* 1985), is an older name than *Wurmiella*. However, the type species, *Y. yaoxianensis* An, 1985, differs in many respects from *O. excavata*, including the characteristic short posterior process on the P_1 element and the varying size of denticles on the assigned ramiform elements; the P_2 element also has a short anterior process and lacks the elevated lobe on the inner side (see An *et al.* 1985, pl. 2, figs 1–7).

Wurmiella amplidentata sp. nov.
Plate 28, figures 1–8

?1983 *Neoprioniodus multiformis* Walliser, 1964; Zhou and Zhai, p. 283, pl. 66, fig. 23 (M element).
?2001 *Ozarkodina adiutricis* Walliser; Li and Qian, p. 96, pl. 1, fig. 2 (P_2 element).

Derivation of name. L., *amplus*, large, plus L., *dentatus*, toothed; in reference to the robust denticles on the P_1 element.

Holotype. Specimen NIGPAS 150096 (Pl. 28, fig. 1); P_1 element.

Type locality and horizon. Huanggexi section, Daguan County, north-east Yunnan Province; uppermost Duluzhai Formation (upper Llandovery), sample TT 1165.

Diagnosis. All elements robust. P_1 element gently arched with relatively few, stout, broad denticles. M element with large cusp and well-developed, denticulate anticusp.

Material. P_1, 2; P_2, 1; M, 2; S_0, 2; S_{3-4}, 1; plus additional material from TT 1165.

Description. P_1 element carminate to gently angulate. Gently arched in lateral view, with even, stout denticles. Cusp of similar size to other denticles and may not be the largest. Anterior process with about five broad denticles, with V-shaped separations, generally increasing a little in size proximally. Posterior process with three or four broad denticles, with wider V-shaped separations than on anterior process. Blade thickened immediately beneath denticle row and may develop into a distinct ledge. Basal cavity with lanaceolate lips, deepest and most widely flared just anterior to cusp, narrowing gradually posteriorly and more abruptly anteriorly to a groove.

P_2 element angulate; cusp very prominent, with sharp anterior and posterior edges. Processes short; anterior process with only one denticle plus an incipient denticle distally, posterior process

with four rather peg-like, discrete denticles that become larger and more posteriorly inclined distally. Cavity flared beneath cusp, with lanceolate lips; inner lip slightly elevated.

M element bipennate, with very robust cusp. Proximally on cusp, axial portion of inner face develops into a broad, low, rounded costa that leads into prominent inner lip to basal cavity. Anticusp well developed as a downward extension of anterior margin of cusp, bearing two or three separated, compressed denticles. Posterior process broken on all specimens, straight to slightly downcurved, bearing large, laterally compressed denticles separated by narrow U-shaped spaces. Basal cavity prominent beneath cusp, continuing as broad groove with central furrow beneath posterior process.

S_0 element alate, with tall cusp with sharp lateral edges. Lateral processes form broad arch of about 160° and bear discrete, rather peg-like denticles. Posterior face of cusp with a broad, low, rounded costa that expands basally over a posteriorly extended basal cavity.

S_{1-2} element not recognized.

S_{3-4} element bipennate with tall, curved cusp. Anterior process directed antero-laterally and downwards, bearing about three discrete denticles with subcircular transverse sections. Posterior process broken on the single specimen, slightly curved, bearing denticles separated by broad, U-shaped spaces; proximal denticle small, the two preserved more distal denticles much larger and somewhat laterally compressed. Basal cavity with slightly flared lips beneath cusp, continuing as broad groove beneath posterior process.

Remarks. The elements compare closely in general morphology to their equivalents in *Wurmiella excavata* (Branson and Mehl, 1933a), although all are more robust with larger, generally more discrete, often somewhat peg-like denticles. The P_1 element is particularly distinctive in its small number of large denticles, and the processes on the P_2 element are much shorter than in typical specimens of *W. excavata*.

There is also a similarity to *Ozarkodina protoexcavata* Cooper, 1975, which should also be re-assigned to *Wurmiella*. S elements of *W. amplidentata* have more discrete, more peg-like denticles, the S_0 element lacks the posterior extension of the cavity, and the P_2 element has shorter processes; the P_1 element figured by Cooper (1975, pl. 3, fig. 1) is broken, but also appears to have more confluent denticles.

Occurrence. Uppermost Daluzhai Formation, Huanggexi section, Daguan County, Yunnan (sample TT 1165).

Wurmiella curta sp. nov.
Plate 28, figures 9–16

Derivation of name. L., *curtus*, short; in reference to the short posterior process on the P_1 element.

Holotype. Specimen NIGPAS 150105 (Pl. 28, fig. 10); P_1 element.

Type locality and horizon. Xuanhe section, Guangyuan County, Sichuan; Shenxuanyi Member, sample Xuanhe 4.

Diagnosis. P_1 element straight with a prominent cusp and a short posterior process; cavity barely flared beneath cusp. Proximal denticle on posterior process typically of similar size to cusp.

Material. P_1, 17; P_2, 16; M, 17; S_0, 18; S_{1-2}, 15; S_{3-4}, 20.

Description. P_1 element. All specimens small. Carminate with basal edge straight or angled slightly upwards anteriorly and posteriorly from basal flare. Cusp prominent, posteriorly inclined. Posterior process less than half the length of anterior process. Anterior process with up to eight small, erect, crowded denticles of variable size; process decreases a little in height distally. Posterior process with up to four posteriorly inclined denticles, decreasing rapidly in size posteriorly; denticle adjacent to cusp commonly of similar size to cusp. Basal cavity only narrowly flared at midlength, barely visible in lateral view. Cusp and all denticles composed of white matter, which extends as a block deeply into blade; lower edge of white matter subparallel to basal edge of element.

P_2 element angulate with a very prominent, very tall cusp, which is slightly twisted and curved inwards, with sharp anterior and posterior edges. Anterior process short, a little more strongly aborally directed than posterior process, bearing up to four compressed, discrete, tall denticles. Posterior process broken on most specimens, but with up to six posteriorly inclined discrete denticles. Cavity flared below cusp, with a stronger, slightly elevated lip on inner side. Cusp and all denticles composed of white matter, which extends deeply into processes. Surfaces of cusp and denticles show subtle microstriae.

M element bipennate; cusp slender, tall, somewhat twisted; cross-section lenticular. Anterior process directed somewhat aborally, with one to three small denticles. Posterior process straight, broken on most specimens, but normally with at least six compressed, confluent denticles. Cavity flared inwards at base of cusp to produce a small lip, then narrowing gradually beneath posterior process. Cusp and all denticles composed of white matter, which extends deeply into processes.

S_0 element alate, with a posteriorly curved, tall cusp, sometimes a little twisted, with sharp lateral edges and flat anterior face; posterior face of cusp with axial ridge, giving a triangular cross-section. Lateral processes form a symmetrical arch of 110–140°, curving very slightly posteriorly, broken on most specimens, but with at least four or five discrete, tall denticles. Posterior ridge on cusp extends into rudimentary posterior process with a narrow, groove-like cavity that extends up cusp face. Cusp and denticles composed of white matter.

S_{1-2} element digyrate, with a tall, twisted, posteriorly inclined cusp. Posterior face of cusp bears an axial ridge of variable prominence. Lateral processes of subequal length, one slightly shorter than the other, forming an arch of 80–110°. Shorter process with three or four discrete denticles, tallest in central part of process; longer process with four or five denticles, tallest in distal part of process, but terminal denticle

smaller. Basal cavity flared posteriorly below cusp to form a small lip or short groove. Cusp and denticles composed of white matter.

S_{3-4} element bipennate with tall, broad, compressed cusp, slightly curved posteriorly. Anterior process curved inwards and slightly downwards, bearing three or four slender denticles. Posterior process broken on all specimens, long, with confluent denticles that increase in size and posterior inclination distally. Basal cavity with slightly flared lips beneath cusp, continuing as narrow groove beneath posterior process. Cusp and denticles composed of white matter.

Remarks. No other *Wurmiella* species are known with such a short posterior process on the P_1 element. In this respect, the species resembles *Yaoxianognathus*, especially the P_1 element of the type species, *Y. yaoxianensis* An, 1985 (see An *et al.* 1985, pl. 2, fig. 6). This perhaps attests to a close relationship between the two genera. *Wurmiella curta* differs from species of *Yaoxianognathus* in the even dentition on all elements and in the characteristically flared cavity on the P_2 element.

Occurrence. Shenxuanyi Member, Xuanhe section, Guangyuan County, Sichuan (sample Xuanhe 4).

Wurmiella puskuensis (Männik, 1994)
Plate 28, figures 17–28

1970 *Ozarkodina* n. sp. B. Pollock *et al.*, p. 757, pl. 113, figs 9–11 (P_1 element).
1994 *Ozarkodina excavata puskuensis* Männik, p. 187, pl. 1, figs 1–10, pl. 2, figs 1–13 (multielement).

Diagnosis. P_1 arched with even-sized triangular denticles, cusp only slightly larger than its neighbours; cavity narrowly flared beneath cusp. Larger specimens of all elements generally have tall, closely spaced denticles. (Emended after Männik 1994, p. 187).

Material. P_1, 88; P_2, 51; M, 26, S_0, 33; S_{1-2}, 31; S_{3-4}, 55; plus additional material from TT samples.

Description. P_1 element angulate, with line joining denticle bases more arched than basal edge; blade high, with free tips of denticles short. Cusp not prominent and may be no larger than adjacent denticles; denticles generally decrease steadily in size distally on both processes. Lateral faces smooth, with no development of lateral thickening. Basal cavity slightly flared on small specimens (e.g. Pl. 28, fig. 19), but becoming less flared and narrower on larger specimens. White matter fills the free denticle tips.

P_2 element angulate, with a prominent, broad, tall cusp, slightly twisted, with sharp anterior and posterior edges. Most specimens broken, but anterior process with four or more discrete denticles; posterior process slightly longer with four or more relatively short denticles, separated by U-shaped spaces. Cavity flared beneath cusp, narrowing abruptly beneath anterior process and more gradually beneath posterior process. Cusp and free tips of denticles composed of white matter.

M element bipennate. Cusp slender, tall, somewhat twisted, with a lenticular cross-section. Shorter process with one to three small denticles. Longer process very gently curved, broken on most specimens, but with at least ten discrete denticles. Cavity flared slightly inwards at base of cusp to produce a small lip, then narrowing gradually beneath posterior process. Cusp and all denticles composed of white matter.

S_0 element alate, with a tall cusp, slightly twisted and gently inclined posteriorly, with sharp lateral edges and flat anterior face; posterior face of cusp rounded or with an axial ridge, giving a triangular cross-section. Lateral processes form a symmetrical arch of up to 140°, curving very slightly posteriorly, broken on most specimens, but with at least five or six discrete denticles. Basal cavity expanded beneath cusp into a narrow groove that extends up posterior face of cusp. Cusp and denticles composed of white matter.

S_{1-2} element digyrate with a tall, slender, twisted, posteriorly inclined cusp. Lateral processes of subequal length, forming an arch of 90–110°. Shorter process with about four discrete, slender denticles, the central two taller than the others; longer process with four or five slender, discrete denticles, normally with a subcircular transverse section. Basal cavity flared posteriorly below cusp to form a small lip. Cusp and denticles composed of white matter.

S_{3-4} element bipennate with a tall cusp, which is lenticular in cross-section and slightly curved posteriorly. Anterior process directed inwards and downwards, bearing up to five slender denticles with subcircular cross-sections. Posterior process broken on most specimens, long, with about eight, discrete slender denticles that increase in size and posterior inclination distally. Basal cavity without flared lips beneath cusp. Cusp and denticles composed of white matter.

Remarks. The elements are similar in general morphology to their equivalents in *Wurmiella excavata* (Branson and Mehl, 1933*a*), but the species is characterized by the high arched blade and short even denticles on the P_1 element. The processes on the P_2 element are much shorter than in typical specimens of *W. excavata*, and the ramiform elements also have shorter processes.

It is the P_1 element that is particularly characteristic of this species, so the diagnosis has been emended to reflect this. As noted by Männik (1994), the denticulation in juvenile specimens is more widely spaced than in larger representatives. The Chinese material differs a little from the type material in the more restricted, narrower cavity of the P_1 element and in the widely spaced denticulation on S_{3-4} specimens of all sizes.

Männik (1994) included in synonymy the specimens referred by Pollock *et al.* (1970, p. 757, figs 5–8) to *Ozarkodina* n. sp. A. However, these specimens are generally straighter, with a more prominent cusp than typical P_1 elements of *W. puskuensis*; some of them also occur in the same sample as an apparent P_2 element of *Wurmiella* with much longer processes (Pollock *et al.* 1970, *Ozarkodina* cf. *O. media*, pl. 113, fig. 15). *Ozarkodina* n. sp. A was included in synonymy with *O. protoexcavata* by Cooper (1975), and this is probably correct.

Occurrence. Lower member, Xiushan Formation, Leijiatun section, Shiqian County, Guizhou; Yangpowan and Shenxuanyi members, Ningqiang Formation, Yushitan section, Ningqiang, Shaanxi.

Wurmiella recava sp. nov.
Plate 29, figures 1–8

Derivation of name. L., *recavus*, concave; in reference to the concave outline of the basal edge of the P_1 element.

EXPLANATION OF PLATE 28

Figs 1–8. *Wurmiella amplidentata* sp. nov. Daluzhai Formation, Huanggexi Section, Daguan County, Yunnan, Sample TT 1165. 1, 150096, P_1 element, lateral view (holotype). 2, 150097, P_1 element, lateral view. 3, 150098, P_2 element, inner lateral view. 4, 150099, M element, inner lateral? view. 5, 150100, M element, inner lateral? view. 6, 150101, S_0 element, posterior view. 7, 150102, S_0 element, posterior view. 8, 150103, S_{3-4} element, inner lateral view.

Figs 9–16. *Wurmiella curta* sp. nov. Shenxuanyi Member, Xuanhe Section, Guangyuan County, Sichuan, Sample Xuanhe 4. 9, 150104, P_1 element, lateral view. 10, 150105, P_1 element, lateral view (holotype). 11, 150106, P_2 element, inner lateral view. 12, 150107, M element, inner lateral? view. 13, 150108, M element, inner lateral? view. 14, 150109, S_{1-2} element, posterior view. 15, 150110, S_0 element, posterior view. 16, 150111, S_{3-4} element, inner lateral view.

Figs 17–28. *Wurmiella puskuensis* (Männik, 1994). Wangjiawan Formation, Yushitan Section, Ningqiang County, Shaanxi, Sample Ningqiang 7. 17, 150112, P_1 element, lateral view. 18, 150113, P_1 element, lateral view. 19, 150114, P_1 element, lateral view. 20, 150115, P_1 element, lateral view. 21, 150116, P_2 element, inner lateral view. 22, 150117, P_2 element, inner lateral view. 23, 150118, S_0 element, posterior view. 24, 150119, M element, inner lateral? view. 25, 150120, S_{1-2} element, posterior view. 26, 150121, S_{3-4} element, inner lateral view. 27, 150122, S_{1-2} element, posterior view. 28, 150123, S_{3-4} element, inner lateral view.

All figures ×60.

PLATE 28

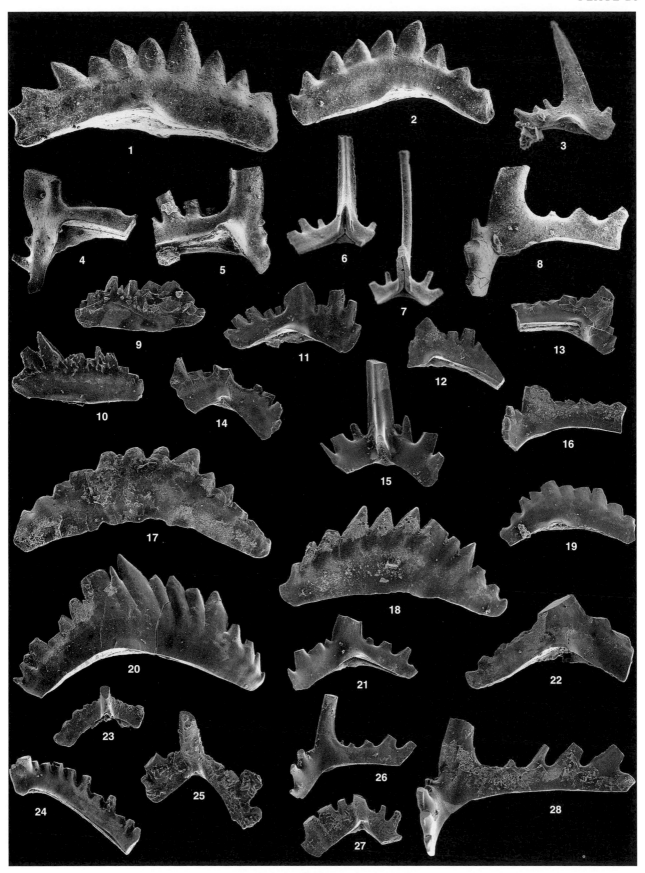

WANG and ALDRIDGE, *Wurmiella*

Holotype. Specimen NIGPAS 150124 (Pl. 29, fig. 1); P_1 element.

Type locality and horizon. Xuanhe section, Gungyuan County, Sichuan; Shenxuanyi Member, sample Xuanhe 3.

Diagnosis. P_1 element short and arched with a fairly prominent cusp and a relatively short posterior process; cavity barely flared beneath cusp, extended under processes as a zone of inverted cavity. P_2 element short, with a large cusp and large, compressed denticles.

Material. P_1, 23; P_2, 7; M, 9; S_0, 2; S_{1-2}, 8; S_{3-4}, 20.

Description. P_1 element angulate. Short and arched, with line joining denticle bases more arched than basal edge. Cusp relatively prominent. Anterior process with 4–8 triangular denticles, those nearest the cusp high and broad, then markedly decreasing in height distally. Posterior process short, with three or four posteriorly inclined denticles, markedly decreasing in size distally. Lateral faces of element smooth, with no development of lateral thickening. Basal cavity barely flared beneath cusp, continuing as a zone of inverted cavity along anterior and posterior processes. White matter fills denticle tips and extends as roots deeply into blade.

P_2 element angulate, no more arched than P_1 element. Cusp prominent, with sharp anterior and posterior edges. Processes short; anterior process with four or five broad, compressed denticles, normally of subequal size but sometimes irregular. Posterior process with two or three broad, compressed denticles that decrease in height distally. Cavity flared beneath cusp, with lanceolate lips; inner lip slightly elevated. White matter fills cusp and denticles and extends into blade.

M element bipennate, with robust, tall, slightly twisted cusp. Lower portion of cusp expanded inwards to form prominent inner lip to basal cavity. Anterior process developed as a downward extension of anterior margin of cusp, bearing up to three separated, compressed denticles. Posterior process straight to slightly downcurved, bearing large, compressed denticles separated by narrow U-shaped spaces. Basal cavity prominent beneath cusp, continuing as broad groove with central furrow beneath posterior process. White matter fills cusp and denticles and extends into blade.

S_0 element alate, with tall cusp with sharp or pinched lateral edges. Lateral processes form a broad arch that becomes steeper distally. Each process with four broad, compressed, discrete denticles. Anterior face of cusp convex in transverse section. Posterior face of cusp with a broad, low, rounded costa that expands basally over a posteriorly extended basal cavity.

S_{1-2} element digyrate, with a tall, twisted, posteriorly inclined cusp with pinched lateral margins. Posterior face of cusp bears a prominent axial ridge. Lateral processes of subequal length, one slightly shorter than the other, both directed downwards to become almost parallel. Shorter process with three or four discrete peg-like denticles, tallest in central part of process; longer process also with three or four denticles, of subequal height, but terminal denticle smaller. Basal cavity flared posteriorly below cusp to form a small lip or a short groove. Cusp and denticles composed of white matter.

S_{3-4} element bipennate with tall, curved cusp. Anterior process directed laterally and downwards, bearing about three discrete denticles with subcircular transverse sections. Two morphologies are apparent, one with the process directed laterally (Pl. 29, fig. 7), the other with the process directed postero-laterally (Pl. 29, fig. 8); these probably represent the S_4 and S_3 positions. Posterior process straight, bearing posteriorly inclined denticles separated by broad, U-shaped spaces; proximal denticles small, the size increasing distally. Basal cavity with very slightly flared lips beneath cusp, continuing as broad groove beneath posterior process. Cusp and denticles composed of white matter.

Remarks. The P_1 element is very distinctive, with its inverted basal cavity and deep white matter; in these respects, it differs from other species of *Wurmiella*, and it may be that this species should be assigned to a separate genus; the P_2 element is also distinctive. However, the rest of the apparatus is similar to that of other *Wurmiella* species, with some elements barely distinguishable from their counterparts in other species; for example, compare the M element with that of *W. amplidentata* (Pl. 28, fig. 5), or the S_{3-4} element with that of *W. puskuensis* (Pl. 28, fig. 28). The closely similar population assigned here to *W.*

EXPLANATION OF PLATE 29

Figs 1–8. *Wurmiella recava* sp. nov. Shenxuanyi Member, Xuanhe Section, Guangyuan County, Sichuan, Sample Xuanhe 3. 1, 150124, P_1 element, lateral view (holotype). 2, 150125, P_1 element, lateral view. 3, 150126, P_2 element, inner lateral view. 4, 150127, M element, inner lateral? view. 5, 150128, S_0 element, posterior view. 6, 150129, S_{1-2} element, posterior view. 7, 150130, S_{3-4} element, inner lateral view. 8, 150131, S_{3-4} element, inner lateral view.

Figs 9–19. *Wurmiella* aff. *recava* sp. nov. Upper member, Xiushan Formation, Leijiatun Section, Shiqian County, Guizhou, Sample Shiqian 20. 9, 150132, P_1 element, lateral view. 10, 150133, P_1 element, lateral view. 11, 150134, P_1 element, lateral view. 12, 150135, P_2 element, inner lateral view. 13, 150136, M element, inner lateral? view. 14, 150137, S_0 element, posterior view. 15, 150138, S_{1-2} element, posterior view. 16, 150139, S_{1-2} element, posterior view. 17, 150140, S_{3-4} element, inner lateral view. 18, 150141, S_0 element, posterior view. 19, 150142, S_{3-4} element, inner lateral view.

All figures ×60.

PLATE 29

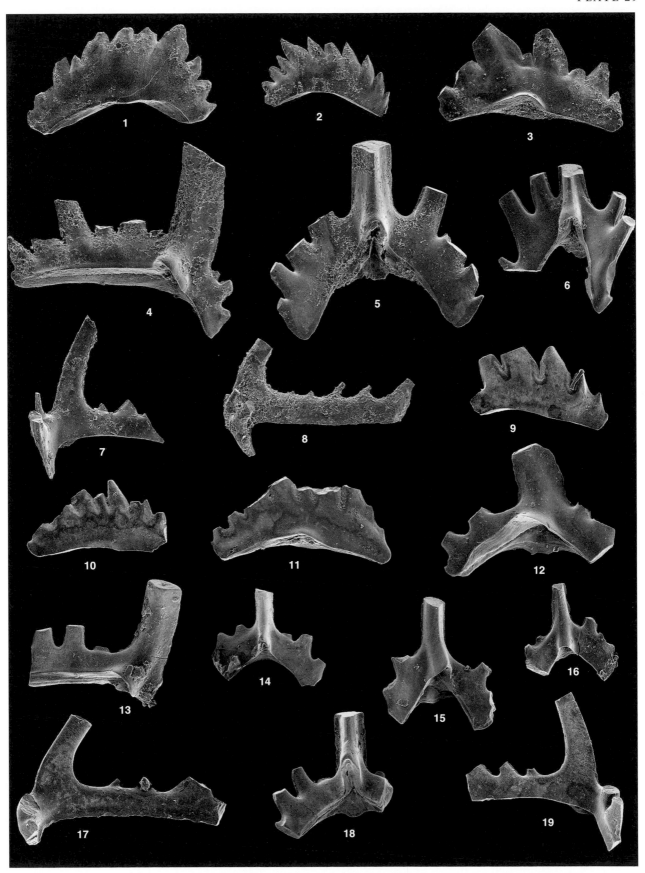

WANG and ALDRIDGE, *Wurmiella*

aff. *recava* has P_1 elements that appear to be intermediate in morphology between those of *W. recava* and those of more typical species of *Wurmiella*.

Occurrence. Shenxuanyi Member, Xuanhe section, Guangyuan County, Sichuan (sample Xuanhe 3).

Wurmiella aff. *recava* sp. nov.
Plate 29, figures 9–18

Material. P_1, 121; P_2, 79; M, 65; S_0, 60; S_{1-2}, 99; S_{3-4}, 130; plus additional material from TT samples.

Remarks. The apparatus of this population is similar to that of *W. recava*. The P_1 elements are comparably short, but are less arched and have only a very narrow zone of inverted basal cavity; the white matter is also not so deeply developed into the denticle roots. Overall, these P_1 specimens are morphologically more similar to typical *Wurmiella* species than are those of *W. recava*. The M and S elements are indistinguishable from those of *W. recava*, but the P_2 element is more strongly arched with more widely spaced dentition. This population may represent a separate species, but it shares so many similarities with *W. recava* that it is not formally separated at present; there is only one good collection of each taxon, and intermediate linking populations may be found.

Occurrence. Upper member, Xiushan Formation, Leijiatun section, Shiqian County, Guizhou (sample Shiqian 20); Shenxuanyi Member, Xuanhe section, Guangyuan County, Sichuan.

Family KOCKELELLIDAE Klapper, 1981

Genus AULACOGNATHUS Mostler, 1967

1967 *Aulacognathus* Mostler, p. 300.
1969 *Neospathognathodus* Nicoll and Rexroad, p. 42.

Type species. *Aulacognathus kuehni* Mostler, 1967, p. 300.

Diagnosis. See Armstrong (1990, p. 62).

Aulacognathus bullatus (Nicoll and Rexroad, 1969)
Plate 27, figures 13–16

*v 1969 *Neospathognathodus bullatus* Nicoll and Rexroad, p. 44, pl. 1, figs 5–7 (P_1 element).
v 1969 *Neospathognathodus ceratoides* Nicoll and Rexroad, p. 46, pl. 1, figs 1–4 (P_2 element).
?1981 *Aulacognathus bullatus* Nicoll et Rexroad, 1968; Zhou and Zhai, p. 269, pl. 65, fig. 8a–b (juvenile P_1 element).
1985 *Aulacognathus bashanensis* Ding and Li, p. 15, pl. 1, figs 16–17 (P_1 element).
1986 *Aulacognathus bullatus*; Nakrem, fig. 8g?, i–j (P_1 element).
1989 *Aulacognathus bullatus* (Nicoll et Rexroad); Yu *in* Jin *et al.*, pl. 3, figs 1a–c, 3a–c; pl. 5, figs 2a–b, 5; pl. 7, figs 2, 6, 9a–b, 11 (P_1 element).
1990 *Aulacognathus bullatus* (Nicoll and Rexroad, 1969); Armstrong, p. 62, pl. 6, figs 1–2, 4–7 (multielement, with synonymy to 1987).
1991b *Aulacognathus bullatus* (Nicoll and Rexroad); McCracken, pl. 3, figs 1–2 (P_2, P_1 elements).

EXPLANATION OF PLATE 30

Figs 1–12. *Ctenognathodus*? *qiannanensis* (Zhou *et al.*, 1981). Lower member, Xiushan Formation, Leijiatun Section, Shiqian County, Guizhou, Sample Shiqian 14B. 1, 150143, P_1 element, lateral view. 2, 150144, P_1 element, lateral view. 3, 150145, P_1 element, lateral view. 4, 150146, P_2 element, inner lateral view. 5, 150147, P_2 element, outer lateral view. 6, 150148, M element, inner lateral view. 7, 150149, S_0 element, posterior view. 8, 150150, S_0 element, posterior view. 9, 150151, S_{3-4} element, inner lateral view. 10, 150152, S_{3-4} element, inner lateral view. 11, 150153, S_{3-4} element, inner lateral view. 12, 150154, S_{1-2} element, posterior view.

Fig. 13. Gen. et sp. indet. A. Shenxuanyi Member, Xuanhe Section, Guangyuan County, Sichuan, Sample Xuanhe 2. 150155, M element, posterior view.

Fig. 14. Gen. et sp. indet. B. Upper member, Xiushan Formation, Leijiatun Section, Shiqian County, Guizhou, Sample Shiqian 17. 150156, Sc? element, inner lateral view.

Figs 15–16. Gen. et sp. indet. C. Upper member, Xiushan Formation, Leijiatun Section, Shiqian County, Guizhou, Sample Shiqian 20. 150157, oral and lateral? views.

Fig. 17. Gen. et sp. indet. D. Shenxuanyi Member, Xuanhe Section, Guangyuan County, Sichuan, Sample Xuanhe 8. 150158, Pa? element, inner lateral view.

Figs 1–12, ×60; figs 13–17, ×80.

PLATE 30

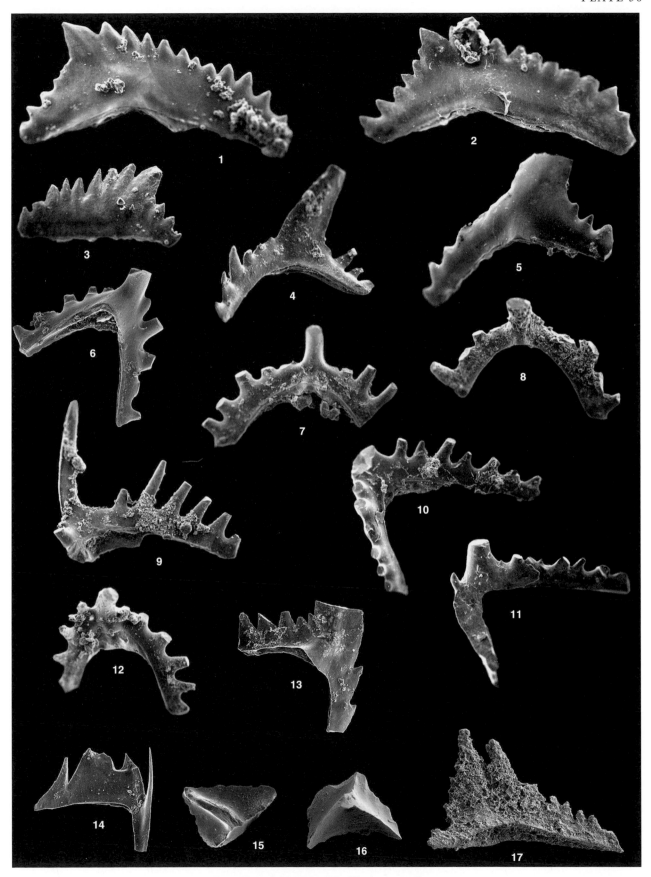

WANG and ALDRIDGE, Silurian conodonts

Diagnosis. 'Apparatus seximembrate; diagnostic stelliscaphate (Pa) element with limited platform development and a long carina which is free to the anterior and posterior of the platform. Inner and outer, anteriorly directed, lateral processes are broad and bear an ornament of irregular coalescing nodes or low sinuous ridges.' (Armstrong 1990, p. 64).

Material. P_1, 5.

Remarks. The P_1 element is variable, but consistently shows a long free blade and broad anteriorly directed inner and outer processes; there is a carina of fused low nodose denticles and a sharply deflected posterior process that may be broadened into a platform. The five incomplete specimens from China differ from those illustrated by Nicoll and Rexroad (1969) primarily in the presence of deep adcarinal grooves separating the lateral processes from the carina, and they also have a longer posterior process. Some other specimens illustrated in the literature share these characteristics (e.g. Over and Chatterton 1987, pl. 3, fig. 1; Armstrong 1990, pl. 6, fig. 1), but they are currently all regarded as falling within the intraspecific range of variation of *A. bullatus.*

Occurrence. Zhanjiawan, Zheng'an County, northern Guizhou Province; from a limestone bed, 0.3–0.4 m thick, resting unconformably on the Hangjiadian Group, and overlain unconformably by the Jiujialu Formation.

Aulacognathus? sp.
Plate 27, figs 17–18

Remarks. Two broken Pa elements from the Daluzhai Formation, Huanggexi section, Yunnan are here tentatively referred to *Aulacognathus.* They bear some resemblance to the broken Pa element of *Apsidognathus oblonguifolius* Melnikov (1999, p. 70, pl. 21, figs 1–4), but in the absence of other elements of the apparatus, the generic attribution is uncertain.

Genus CTENOGNATHODUS Fay, 1959

1856 *Ctenognathus* Pander, p. 32.
1959 *Ctenognathodus* Fay, p. 195.

Type species. Ctenognathus murchisoni Pander, 1856, by monotypy.

Remarks. As noted by Viira and Einasto (2003, p. 228) the P_1 element of *Ctenognathodus* resembles the homologous element of *Ozarkodina*, whereas the M and S elements are similar to those of *Oulodus.*

Ctenognathodus? *qiannanensis* (Zhou, Zhai and Xian, 1981)
Plate 30, figs 1–12

  *1981 *Ozarkodina qiannanensis* Zhou et al., p. 136, pl. 2, figs 24–25 (P_1 element).
  1983 *Ozarkodina qiannanensis* Zhou, Zhai et Xian, 1981; Zhou and Zhai, p. 289, pl. 67, fig. 13 (P_1 element).
?1983 *Ozarkodina alisonae* Aldridge, 1972; Zhou and Zhai, p. 287, pl. 67, fig. 10 (P_2 element).
 v 1996 *Ctenognathodus*? *qiannanensis* (Zhou et al.); Wang and Aldridge, pl. 3, figs 2–3 (P_2, P_1 elements).
 v 2002 *Ctenognathodus*? *qiannanensis* (Zhou et al.); Aldridge and Wang, fig. 64B–C (copy of Wang and Aldridge 1996, pl. 3, figs 2–3) (P_2, P_1 elements).

Diagnosis. P_1 element carminate to angulate with a short posterior process and a higher, long, nearly straight anterior process. P_2 element angulate with a wide, tall, robust cusp and slightly twisted posterior process. M and S elements all with widely spaced peg-like denticles and slightly twisted processes.

Material. P_1, 148; P_2, 64; M, 50; S_0, 25; S_{1-2}, 43; S_{3-4}, 101.

Description. P_1 element. Juvenile specimens carminate; adults gently angulate. Blade thick, robust, curved gently inwards in posterior quarter of unit. Cusp situated one quarter of distance from posterior tip; broad, laterally compressed, nearly triangular in lateral view and inclined posteriorly. Posterior process short with three to six low triangular denticles, fused at their bases, posteriorly inclined; height of posterior process decreases markedly distally to form a steep upper margin. Anterior process two to three times longer than posterior process, bearing eight to 12 closely spaced denticles; proximal denticle joins cusp at higher level than proximal denticle on posterior process. Height of anterior process decreases only a little anteriorly in adult specimens; lower edge straight. Basal cavity flared beneath cusp; outer lip more extensive than inner lip.

  P_2 element strongly angulate, slightly bowed. Cusp very tall, laterally compressed with sharp anterior and posterior edges, posteriorly inclined and curved slightly inwards. Posterior process low and short, slightly twisted, with about three discrete denticles. Anterior process about twice length of posterior process, with about six discrete denticles. Basal cavity moderately expanded beneath cusp, forming a small outer lip; cavity extending as narrow groove under anterior process and as broader groove beneath posterior process.

  M element bipennate, the two processes forming an acute arch. Cusp stout, inwardly curved, with a subcircular cross-section;

anterior and posterior margins rounded. Posterior process long and strongly downcurved, sometimes slightly twisted, bearing four or five discrete, peg-like denticles that increase in size distally. Aboral surface of posterior process broad, with a narrow axial groove. Anterior process of variable length, directed downwards, bearing three to five peg-like denticles of subequal size. Basal cavity moderately deep with an inwardly flared lip beneath cusp.

S_0 element alate, with lateral processes forming a broadly curved, almost semi-circular arch. Cusp tall, posteriorly curved, ovoid in cross-section with slightly sharp lateral edges. Each lateral process bears four to eight subequal, widely spaced, tall peg-like denticles with biconvex cross-sections. Basal cavity with small posteriorly flared lip beneath cusp, extending as a groove beneath each process.

$S_{1–2}$ element breviform digyrate, with a tall, strongly posteriorly curved cusp, subcircular to ovoid in cross-section. Shorter lateral process directed strongly downwards and slightly posteriorly, bearing four or five widely spaced peg-like denticles. Longer process curved more gently downwards, straight or very slightly twisted, supporting five or six widely spaced peg-like denticles. Basal cavity moderately deep with a posteriorly flared lip beneath cusp; basal body present on some specimens.

$S_{3–4}$ element bipennate with a tall, gently posteriorly curved cusp, biconvex in cross-section. Posterior process long, straight or slightly twisted outwardly, bearing four to seven tall, widely spaced peg-like denticles with biconvex or subcircular cross-sections. Antero-lateral process shorter, extends downwards and very slightly posteriorly, bearing four to six discrete, widely spaced denticles with subcircular cross-sections. Basal cavity deepest and widest, but not flared, beneath cusp and extending as a groove beneath the processes.

Remarks. Specimens are dark brown, and the extent of white matter cannot be determined. The P_1 and P_2 elements of this apparatus are distinctive, but bear a broad resemblance to the homologous elements in the apparatus of *Ctenognathodus jeppssoni* Viira and Einasto, 2003 (p. 228, pl. 1, figs 1–6, 8, 11–12). The P_1 element of *C. jeppssoni* has far fewer, more robust denticles and the P_2 element, though similarly slightly twisted, has much taller, more discrete denticles. The P_1 element of the type species, *C. murchisoni*, is less arched and has a higher posterior process than that of *C.? qiannanensis* (see Viira 1982, p. 75, fig. 5.6, pl. 7, figs 1–15, pl. 8, figs 1–2, 4).

Identification of the M and S elements of *C.? qiannanensis* is complicated by the fact that all associated elements in the same sample are *Oulodus*-like, including specimens here assigned to *Oulodus tripus* sp. nov. Assignment of ramiform specimens to *Ctenognathodus?* and *Oulodus tripus* is somewhat equivocal; we have illustrated a suite of elements with consistent denticulation, and long processes that we consider are the most likely to represent the ramiform array of *Ctenognathodus?* However, we have distinguished no $S_{3–4}$ element for

O. tripus, and it is likely that some specimens of this element have been misassigned to *C.? qiannanensis*. Final consideration of whether *qiannanensis* should be referred to *Ctenognathodus* or to a new genus awaits full understanding of the entire apparatus.

Occurrence. Lower member, Xiushan Formation, Leijiatun section, Shiqian County, Guizhou. Also reported from the Shanggaozhaitian Group, Wudang, Guizhou and from Guiding, Guizhou (Zhou *et al.* 1981).

Order UNKNOWN
Family UNKNOWN

Genus MULTICOSTATUS Ding and Li, 1985

Type species. Multicostatus dazhubaiensis Ding and Li, 1985, p. 17, pl. 1, figs 9–10, by monotypy.

Diagnosis. A conical cusp, with the apex rounded and curving posteriorly. There are several longitudinal ridges on the surface, which converge posteriorly at the tip. Basal cavity medium. (Ding and Li 1985, p. 17, translated by Ma Xiao-ya).

Remarks. The status of the genus *Multicostatus* is uncertain. There are similarities to specimens of *Tuberocostadontus*, and the two taxa may represent members of the same multielement apparatus (see Wang and Aldridge 1998). Fordham (1991) included both genera in his synonymy of *Tuberocostadontus*, but a complete suite of elements was recognized by Bischoff (1986) for *Tuberocostadontus* (as *Pyrsognathus*) and is also identified here (see above). The presumed S elements of *Multicostatus* share features that are lacking in *Tuberocostadontus* elements, especially the coalescence of the longitudinal ridges into a reticulate apical region. *Multicostatus* is, therefore, currently retained as a separate genus, but the nature of any additional elements of the apparatus is highly equivocal.

Multicostatus dazhubaensis Ding and Li, 1985
Plate 13, figures 11–14, 17–20, ?15–16, ?21

1985 Multicostatus dazhubaensis Ding and Li, p. 17, pl. 1, figs 9–10.

Diagnosis. As for genus.

Material. 56 specimens, plus additional material from TT samples.

Remarks. Specimens we refer to this species range from tall cones to squat cones with an expanded base; they may represent at least three separate elements within a multielement apparatus.

The holotype is a relatively tall, robust cone, curving gently posteriorly, with more strongly curved costae developed on each lateral face; one lateral face is slightly expanded basally to form an incipient lateral process. Specimens in the material we have examined include cones with a very similar shape (Pl. 13, figs 11–12), but the costae are commonly more irregularly developed, with short costae interspersed between the longer ones near the base; there is a narrow, acostate margin around the base. At the blunt apex of the unit the costae converge into a reticulated area. One specimen (Pl. 13, figs 13–14) is extended on one side into a basal lip, with two short, asymmetrically developed processes, one incipient and the other longer and more laterally directed with a slightly serrate upper surface; this specimen broadly resembles the holotype of *T. shiqianensis* in morphology (Zhou *et al.* 1981, p. 139, fig. 8, pl. 1, fig. 52), but there is no evidence of a reticulate tip on the illustrations of the latter.

Shorter cones also occur, all with the characteristic ridges and apical reticulation; these include specimens with an expanded base with a narrow acostate margin (Pl. 13, figs 17–18), and less expanded specimens with a broader smooth basal zone (Pl. 13, figs 19–20). We tentatively view this suite of conical elements as representing the S series of the apparatus.

A squat, more triangular specimen with just a few short costae and a slightly serrate anterior edge (Pl. 13, fig. 21) occurs in association with the other elements and may be part of the same apparatus; it resembles specimens assigned to the Pb position of *Tuberocostadontus* and may be from that genus, but it could also be from *Multicostatus* and attest to a close relationship between the two genera.

A single specimen from sample Xuanhe 3 (Pl. 13, figs 15–16) has a squat coniform morphology with deep, nodose rides developed above a smooth basal margin. The basic shape is similar to that of squat cones of *Multicostatus*, and it may belong to the same apparatus, to another species of the same genus, or to a related genus. It is possible that this specimen may not even be a conodont element, but a dermal denticle of a fish.

Occurrence. Upper member, Xiushan Formation, Leijiatun section, Shiqian County, Guizhou; Yangpowan Member, Ningqiang Formation, Yushitan section, Ningqiang, Shaanxi; Shenxuanyi Member, Xuanhe section, Guangyuan County, Sichuan.

Acknowledgements. This work arises from a co-operative project entitled 'Transhemisphere Telychian: a biostratigraphical experiment' (TT programme) between the Chinese Academy of Sciences and the Royal Society; we express our gratitude to the funding agencies and to Prof. Chen Xu and Prof. C. H. Holland, leaders of the TT programme. Financial support was kindly also provided by the Royal Society for Wang Cheng-yuan to visit Leicester for 3 months to progress this work; Wang also thanks NSFC for project 40839910. The Leijiatun, Xuanhe and Ningqiang sections were measured by Chen Xu, Rong Jia-yu, Wang Cheng-yuan, Geng Liang-yu, Chen Ting-en, Xu Jun-tao, Wu Hong-ji and Deng Zhan-qiu. After the TT programme, Chen-Xu, Rong Jia-yu, Wang Cheng-yuan, Geng Liang-yu and Deng Zhan-qiu measured and sampled the Huanggexi section in Yunnan and the Erlangshan section in Sichuan. Processing and SEM facilities were provided by the Nanjing Institute of Geology and Palaeontology and by the University of Leicester, where the help of Andrew Swift and Rod Branson is acknowledged. Lisa Barber and Tim Lambert produced the finished versions of the line diagrams. Drs Zhen Yong-yi, B. G. Fordham and Ma Xiao-ya are thanked for providing English translations of Chinese literature, and Dr S. Curtis is thanked for the photographs in Text-figure 13. The article was improved following thorough reviews by Drs Peep Männik and Zhang Shun-xin.

Editor. Svend Stouge

REFERENCES

ALDRIDGE, R. J. 1972. Llandovery conodonts from the Welsh Borderland. *Bulletin of the British Museum Natural History Geology*, **22**, 125–231.
—— 1974. An *amorphognathoides* Zone conodont fauna from the Silurian of the Ringerike area, south Norway. *Norsk Geologisk Tidsskrift*, **54**, 295–303.
—— 1975. The stratigraphic distribution of conodonts in the British Silurian. *Journal of the Geological Society, London*, **131**, 607–618.
—— 1979. An upper Llandovery conodont fauna from Peary Land, eastern North Greenland. *Rapport Grønlands Geologiske Undersøgelse*, **91**, 7–23.
—— 1982. A fused cluster of coniform conodont elements from the Late Ordovician of Washington Land, Western North Greenland. *Palaeontology*, **25**, 425–430.
—— 1985. Conodonts of the Silurian System from the British Isles. 68–94. *In* HIGGINS, A. C. and AUSTIN, R. L. (eds). *A stratigraphical index of conodonts*. Ellis Horwood, Chichester, 263 pp.
—— 2002. Conodonts from the Skomer Volcanic Group (Llandovery Series, Silurian) of Pembrokeshire, Wales. *Special Papers in Palaeontology*, **67**, 15–28.
—— and SCHÖNLAUB, H. P. 1989. Conodonts. 274–279. *In* HOLLAND, C. H. and BASSETT, M. G. (eds). *A global standard for the Silurian System*. National Museum of Wales, Cardiff, 325 pp.
—— and SMITH, M. P. 1993. Conodonta. 563–572. *In* BENTON, M. J. (ed.). *The fossil record 2*. Chapman and Hall, London, xvii + 845 pp.
—— and WANG CHENG-YUAN 2002. Conodonts. 83–94. *In* HOLLAND, C. H. and BASSETT, M. G. (eds). *Telychian rocks of the British Isles and China (Silurian, Llandovery Series)*. National Museums and Galleries of Wales, Geological Series No. 21, 210 pp.
—— PURNELL, M. A., GABBOTT, S. E. and THERON, J. N. 1995. The apparatus architecture and function of *Promis-*

sum pulchrum Kovács-Endrödy (Conodonta, Upper Ordovician) and the prioniodontid plan. *Philosophical Transactions of the Royal Society of London, Series B*, **347**, 275–291.

AN TAI-XIANG 1987. *The Lower Paleozoic conodonts of South China*. Beijing University Publishing House, 238 pp. [In Chinese].

—— and ZHENG ZHAO-CHANG 1990. *The conodonts of the marginal area around the Ordos Basin, North China*. Science Press, Beijing, 201 pp. [In Chinese with English abstract].

—— ZHANG AN-TAI and XU JIAN-MIN 1985. Ordovician conodonts from Yaoxian and Fuping, Shaanxi and their stratigraphic significance. *Acta Geologica Sinica*, **59**, 97–108. [In Chinese with English abstract].

ARMSTRONG, H. A. 1990. Conodonts from the Upper Ordovician – Lower Silurian carbonate platform of North Greenland. *Bulletin Grønlands Geologiske Undersøgelse*, **159**, 1–151.

AUSTIN, R. L. (ed.) 1987. *Conodonts: investigative techniques and applications*. Ellis Horwood Limited, Chichester, for the British Micropalaeontological Society, 422 pp.

BARNES, C. R., KENNEDY, D. J., McCRACKEN, A. D., NOWLAN, G. S. and TARRANT, G. A. 1979. The structure and evolution of Ordovician conodont apparatuses. *Lethaia*, **12**, 125–151.

BARRICK, J. E. 1977. Multielement simple cone conodonts from the Clarita Formation (Silurian), Arbuckle Mountains, Oklahoma. *Geologica et Palaeontologica*, **11**, 47–68.

—— and KLAPPER, G. 1976. Multielement Silurian (late Llandoverian – Wenlockian) conodonts of the Clarita Formation, Arbuckle Mountains, Oklahoma, and phylogeny of *Kockelella*. *Geologica et Palaeontologica*, **10**, 59–100.

BASSLER, R. S. 1925. Classification and stratigraphic use of the conodonts. *Geological Society of America Bulletin*, **36**, 218–220.

BISCHOFF, G. C. O. 1986. Early and Middle Silurian conodonts from midwestern New South Wales. *Courier Forschungsinstitut Senckenberg*, **89**, 1–337.

—— 1997. *Ansella mischa* n. sp. (Conodonta) from late Llandoverian and early Wenlockian strata of Midwestern New South Wales. *Neues Jahrbuch für Geologie und Paläontologie Monatshefte*, **1997**, 477–488.

—— and SANNEMANN, D. 1958. Unterdevonische Conodonten aus dem Frankenwald. *Notizblatt des hessisches Landesamt für Bodenforschung zu Wisbaden*, **86**, 87–110.

BRANSON, E. B. and BRANSON, C. C. 1947. Lower Silurian conodonts from Kentucky. *Journal of Paleontology*, **21**, 549–556.

—— and MEHL, M. G. 1933*a*. Conodonts from the Bainbridge (Silurian) of Missouri. *University of Missouri Studies*, **8**, 39–52.

—— —— 1933*b*. Conodonts from the Joachim (Middle Ordovician) of Missouri. *University of Missouri Studies*, **8**, 77–100.

—— —— 1933*c*. Conodonts from the Plattin (Middle Ordovician) of Missouri. *University of Missouri Studies*, **8**, 101–119.

—— —— 1941. New and little known Carboniferous conodont genera. *Journal of Paleontology*, **15**, 97–106.

BRAZAUSKAS, A. Z. 1983. Conodont zones of Lithuanian Llandovery facies. *Nauchnye Trudy Vysshih Uchebnyh Zavedenij Litovskoj SSR, Geologija*, **4**, 41–66. [In Russian, with English and Lithuanian summaries].

CHAUFF, K. M. and PRICE, R. C. 1980. *Mitrellataxis*, a new multielement genus of Late Devonian conodont. *Micropaleontology*, **26**, 177–188.

CHEN XU and LIN YAO-KUN 1978. Lower Silurian graptolites from Tongxi, northern Guizhou. *Memoirs of the Nanjing Institute of Geology and Palaeontology*, **12**, 1–106.

—— and RONG JIA-YU (eds) 1996. *Telychian (Llandovery) of the Yangtze region and its correlation with British Isles*. Science Press, Beijing. 162 pp. [In Chinese with English abstract].

—— —— WANG CHENG-YUAN, GENG LIANG-YU, DENG ZHAN-QIU, WU HONG-JI, XU JUN-YAO and CHEN TING-EN 2002. Telychian rocks in the Yangtze region. 13–43. *In* HOLLAND, C. H. and BASSETT, M. G. (eds) 2002. *Telychian rocks of the British Isles and China (Silurian, Llandovery Series): an experiment to test precision in stratigraphy*. National Museum of Wales Geological Series No. 21, Cardiff, 210 pp.

CHENG YU-QI (ed.) 1994. *Outline of Regional Geology of China*. Geological Publishing House, Beijing. 517 pp. [In Chinese].

COOPER, B. J. 1975. Multielement conodonts from the Brassfield Limestone (Silurian) of southern Ohio. *Journal of Paleontology*, **49**, 984–1008.

—— 1976. Multielement conodonts from the Saint Clair Limestone (Silurian) of southern Illinois. *Journal of Paleontology*, **50**, 205–217.

—— 1977. Toward a familial classification of Silurian conodonts. *Journal of Paleontology*, **51**, 1057–1071.

CORRADINI, C. 2001. Il genere *Pseudooneotodus* Drygant (Conodonta) nel Siluriano e Devoniano Inferiore della Sardegna. *Giornale di Geologia*, **62**, Suppl., 23–29.

—— 2008 (for 2007). The conodont genus *Pseudooneotodus* Drygant from the Silurian and Lower Devonian of Sardinia and the Carnic Alps (Italy). *Bolletino della Società Paleontologica Italiana*, **46**, 139–148.

DING MEI-HUA and LI YAO-QUAN 1985. Silurian conodonts of Ningqiang area, Shaanxi Province and their biostratigraphic significance. *Earth Science – Journal of Wuhan College of Geology*, **10**, 9–20. [In Chinese with English abstract].

DONOGHUE, P. C. J. 1998. Growth and patterning in the conodont skeleton. *Philosophical Transactions of the Royal Society of London, Series B*, **353**, 633–666.

—— FOREY, P. L. and ALDRIDGE, R. J. 2000. Conodont affinity and chordate phylogeny. *Biological Reviews*, **75**, 191–251.

—— PURNELL, M. A., ALDRIDGE, R. J. and ZHANG SHUN-XIN 2008. The interrelationships of 'complex' conodonts (Vertebrata). *Journal of Systematic Palaeontology*, **6**, 119–153.

DRYGANT, D. M. 1974. Simple conodonts from the Silurian and lowermost Devonian. *Palaeontologischeskii Sbornik*, **10**, 64–70.

DUMOULIN, J. A., BRADLEY, D. C., HARRIS, A. G. and REPETSKI, J. E. 1997. Lower Paleozoic deep-water facies of the Medfra area, central Alaska. 73–103. *In* KELLY, K. D. (ed.). *Geologic studies in Alaska by the U. S.*

Geological Survey, 1997. U. S. Geological Survey Professional Paper, **1614**, 160 pp.

DZIK, J. 1976. Remarks on the evolution of Ordovician conodonts. *Acta Palaeontologica Polonica*, **21**, 395–455.

—— 1991. Evolution of the oral apparatuses in the conodont chordates. *Acta Palaoentologica Polonica*, **36**, 265–323.

—— 2006. The Famennian 'golden age' of conodonts and ammonoids in the Polish part of the Variscan Sea. *Palaeontologia Polonica*, **63**, 1–359.

ETHINGTON, R. L. 1959. Conodonts of the Ordovician Galena Formation. *Journal of Paleontology*, **33**, 257–292.

FÅHRAEUS, L. E. and BARNES, C. R. 1981. Conodonts from the Becscie and Gun River formations (Lower Silurian) of Anticosti Island, Québec. 165–172. In LESPÉRANCE, P. J. (ed.). *IUGS Subcommission on Silurian Stratigraphy (Ordovician–Silurian Boundary Working Group) Field Meeting. Anticosti-Gaspé, Québec, 1981, Volume 2, Stratigraphy and paleontology.* Département de Géologie, Université de Montréal, v + 321 pp.

FANG RUN-SEN 1979. *Silurian of Qujing.* Selected papers of Stratigraphy and Palaeontology of the second conference of the Geological Society of Yunnan [not formally published].

—— JIANG NEN-REN, FAN JIAN-CAI, CAO REN-GUAN, LI DAI-YUN, et al. 1985. *The Middle Silurian and Early Devonian stratigraphy and palaeontology in Qujing District, Yunnan.* Yunnan People's Republic of China, Kunming, 171 pp. [In Chinese with English abstract].

FAY, R. O. 1959. Generic and subgeneric homonyms of conodonts. *Journal of Paleontology*, **33**, 195–196.

FORDHAM, B. G. 1991. A literature-based phylogeny and classification of Silurian conodonts. *Palaeontographica Abteil A*, **217**, 1–136.

GAGIEV, M. H. 1979. Conodonts from the Devonian/Carboniferous boundary deposits of the Omolon Massif. Field Excursion Guidebook for Tour 9, Biostratigraphy and fauna of Devonian-Carboniferous boundary deposits, 14th Pacific Science Congress, Khabarocsk, USSR, August 1979, Supplement 2, 3–104. [In Russian, with English diagnoses of new genera and species].

GE ZHI-ZHOU and YU CHANG-MING 1974. Silurian corals. 165–173. In NANJING INSTITUTE OF GEOLOGY AND PALAEONTOLOGY, ACADEMIA SINICA (ed.). *Handbook of stratigraphy and palaeontology of Southwest China.* Science Press, Beijing, 454 pp. [In Chinese].

—— RONG JIA-YU, YANG XUE-CHANG, LIU GENG-WU, NI YU-NAN, DONG DE-YUAN and WU HONG-JI 1977. The materials of ten sections of Silurian system in Southwestern regions of China. *Stratigraphy and Palaeontology*, **8**, 92–111. [In Chinese].

—— —— —— —— —— —— 1979. Silurian in Southwestern regions of China. 155–220. In NANJING INSTITUTE OF GEOLOGY AND PALAEONTOLOGY, ACADEMIA SINICA (ed.). *Carbonatite Biostratigraphy in Southwerstern regions of China.* Science Press, Beijing, 336 pp. [In Chinese].

GENG LIANG-YU 1990. Chitinozoa near Aeronian-Telychian boundary at Leijiatun of Shiqian, N. Guizhou. *Acta Palaeonto-logica Sinica*, **29**, 623–636. [In Chinese with English summary].

GIRARD, C. and WEYANT, M. 1996. Conodontes Siluriens de l'Ile Hoved (Archipel Arctique Canadien). *Revue de Micropaléontologie*, **39**, 53–66.

GLENISTER, A. T. 1957. The conodonts of the Ordovician Maquoketa Formation in Iowa. *Journal of Paleontology*, **31**, 715–736.

GUO WEN-KUI and HUANG SHAO-XIAN 1942. Geological mineral resources of Yanjun-Daguan-Yiliang areas in Yunnan. *Southwest Mineral Prospecting Bureau of Resource Committee, Provisional Reports*, **17**, 1–13. [In Chinese].

HASS, W. H. 1959. Conodonts from the Chappel Limestone of Texas. *U. S. Geological Survey Professional Paper* **294-J**, 363–399.

HE YONG-YING 1983. Discovery of Late Silurian conodonts in North Qilian Mts. *Regional Geology of China*, **6**, 138–140. [In Chinese].

HE YUAN-XIANG and QIAN YONG-ZHEN 2000. The Silurian–Devonian boundary in Baizitian, Yanbian, Sichuan and its geological implications. *Sedimentary Geology and Tethyan Geology*, **20**, 98–112. [In Chinese with English abstract].

HOLLAND, C. H. and BASSETT, M. G. (eds) 2002. *Telychian rocks of the British Isles and China (Silurian, Llandovery Series): an experiment to test precision in stratigraphy.* National Museum of Wales Geological Series No. 21, Cardiff, 210 pp.

JEPPSSON, L. 1972. Some Silurian conodont apparatuses and possible conodont dimorphism. *Geologica et Palaeontologica*, **6**, 51–69.

—— 1975 [for 1974]. Aspects of late Silurian conodonts. *Fossils and Strata*, **6**, 54 pp.

—— 1979. Conodonts. 225–248. In JAANUSSON, V., LAUFELD, S. and SKOGLUND, R. (eds). Lower Wenlock faunal and floral dynamics- Vattenfallet section, Gotland. *Sveriges Geologiska Undersökning Avdhandlingar och Uppsatser*, **C762**, 294 pp.

—— 1983. Simple cone studies: some provocative thoughts. *Fossils and Strata*, **15**, 86.

—— 1988. Conodont biostratigraphy of the Silurian-Devonian boundary stratotype at Klonk, Czechoslovakia. *Geologica et Palaeontologica*, **22**, 21–31.

—— 1989. Latest Silurian conodonts from Klonk, Czechoslovakia. *Geologica et Palaeontologica*, **23**, 21–37.

—— 1997. A new latest Telychian, Sheinwoodian and early Homerian (Early Silurian) standard conodont zonation. *Transactions of the Royal Society of Edinburgh: Earth Sciences*, **88**, 91–114.

—— 1998. Silurian oceanic events: a summary of general characteristics. 239–257. In LANDING, E. and JOHNSON, M. E. (eds). *Silurian cycles: linkages of dynamic stratigraphy with atmospheric, oceanic, and tectonic changes.* New York State Museum Bulletin, **491**, 327 pp.

—— FREDHOLM, D. and MATTIASSON, B. 1985. Acetic acid and phosphatic fossils – a warning. *Journal of Paleontology*, **59**, 952–956.

JIANG WU, ZHANG FANG, ZHOU YU-YIN, XIONG JIAN-FEI, DAI JIN-YIE and ZHONG DUAN 1986. Silurian conodonts. 105–119. In *Conodonts – palaeontology.*

Southwestern Petroleum Institute, Sichuan, 264 pp. [In Chinese].

JIN CHUN-TAI, QIAN YONG-ZHEN and WANG JI-LI 2005. Silurian conodont succession and chronostratigraphy of the Baizitian region in Yanbian, Panzhihua, Sichuan. *Journal of Stratigraphy*, **29**, 281–294. [In Chinese with English abstract].

—— WAN ZHENG-QUAN, YE SHAO-HUA, CHEN JI-RONG and QIAN YONG-ZHEN 1992. *The Silurian System in Guangyuan, Sichuan and Ningqiang, Shaanxi.* Chengdu Sciences and Technology University Press, Chengdu, 97 pp. [In Chinese with English abstract].

—— YE SHAO-HUA, JIANG XIN-SHENG, LI YU-WEN, YU HONG-JIN, HE YUAN-XIANG, YI YONG-EN and PAN YUN-TANG 1989. Silurian stratigraphy and palaeontology in Erlangshan District, Sichuan. *Bulletin of the Chengdu Institute of Geology and Mineral Resources, Chinese Academy of Geological Sciences*, **11**, 224 pp. [In Chinese with English abstract].

KHODALEVICH, A. N. and TSCHERNICH, V. V. 1973. Novoe podsemeystvo Belodellinae (Konodonty). *Trudy Sverdlovskogo Gornogo Instituta*, **93**, 42–47. [In Russian].

KLAPPER, G. 1981. Family Distomodontidae Klapper, new. W137. *In* ROBISON, R. A. (ed.). *Treatise on Invertebrate Paleontology, Part W, Miscellanea, Supplement 2, Conodonta.* Geological Society of America, Boulder, Colorado and the University of Kansas Press, Lawrence, Kansas, xxviii + 202 pp.

KOZUR, H. and MOSTLER, H. 1970. Neue Conodonten aus der Trias. *Berichte des Naturwissenschaftlich-medizinischen Vereins in Innsbruck*, **58**, 429–464.

KŘÍŽ, J. 1989. The Přídolí Series in the Prague Basin (Barrandian area, Bohemia). 90–100. *In* HOLLAND, C. H. and BASSETT, M. G. (eds). *A global standard for the Silurian System.* National Museum of Wales Geological Series, **9**, Cardiff, 325 pp.

LAN CHAO-HUA 1979. Silurian sequence of Daguan area, NE Yunnan. *Selected Proceedings of Stratigraphy and Palaeontology*, 71–88. Geological Society of Yunnan Province. [In Chinese].

LI JIN-SENG 1987. Late Silurian – Devonian conodonts from Luqu-Tewo Region, West Qingling Mountains, China. 357–378. *In* XI'AN INSTITUTE OF GEOLOGY AND MINERAL RESOURCES, NANJING INSTITUTE OF GEOLOGY AND PALAEONTOLOGY, ACADEMIA SINICA (eds). *Late Silurian – Devonian strata and fossils from Luqu-Tewo area of West Qingling Mountains, China.* Nanjing University Press, 450 pp. [In Chinese with English abstract].

LI ZHONG-XIONG and QIAN YONG-ZHEN 2001. Recent progress in the research on the Silurian conodonts from the western margin of the Yangtze Platform. *Sedimentary Geology and Tethyan Geology*, **21**, 87–101. [In Chinese with English abstract].

LIN BAO-YU 1979. The Silurian System of China. *Acta Geologica Sinica*, **53**, 173–191. [In Chinese with English abstract].

—— 1983. New developments in conodont biostratigraphy of the Silurian of China. *Fossils and Strata*, **15**, 145–148.

—— 1984. *The Silurian System of China; Stratigraphy of China, 6.* Geological Publishing House, Beijing, 245 pp., 9 pls. [In Chinese].

—— 1986. Progress and prospects of the research of Silurian conodont biostratigraphy in China. *Geology of China*, **10**, 31–32. [In Chinese].

—— 1991. New progress of Silurian study in Yangtze Platform. *Geology of China*, **1**, 16–17. [In Chinese].

—— and QIU HONG-RONG 1983. The Silurian System in Xizang (Tibet). *Contribution to the Geology of the Qinghai-Xizang (Tibet) Plateau*, **8**, 15–28. [In Chinese with English abstract].

—— —— 1985. New progress of the research of Palaeozoic stratigraphy and palaeontology in Tibet. *Geology of China*, **1985**, 8. [In Chinese].

LINDSTRÖM, M. 1970. A suprageneric taxonomy of the conodonts. *Lethaia*, **3**, 427–455.

LINK, A. G. and DRUCE, E. C. 1972. Ludlovian and Gedinnian conodont stratigraphy of the Yass Basin, New South Wales. *Bulletin of the Bureau of Mineral Resources, Geology and Geophysics, Australia*, **134**, 1–136.

LIU DIAN-SHENG, YANG JI-KAI and FU YING-QI 1993. The feature of Silurian conodonts biota in the northern sections of Longmen Mountain, Sichuan. *Journal of Central-South Institute of Mining and Metallurgy*, **24**, 573–574. [In Chinese with English abstract].

LIU JIA-DUO, ZHANG CHENG-JIANG, LIU XIAN-FAN, LI YOU-GUO, YANG ZHENG-XI and WU DE-CHAO 2004. *Mineralization, regulation and exploration evaluation in southwest margin of Yangtze Platform.* Geological Publishing House, Beijing. 204 pp. [In Chinese].

MABILLARD, J. E. and ALDRIDGE, R. J. 1983. Conodonts from the Coralliferous Group (Silurian) of Marloes Bay, South-West Dyfed, Wales. *Geologica et Palaeontologica*, **17**, 29–36.

—— —— 1985. Microfossil distribution across the base of the Wenlock Series in the type area. *Palaeontology*, **28**, 89–100.

MÄNNIK, P. 1983. Silurian conodonts from Severnaya Zemlya. *Fossils and Strata*, **15**, 111–119.

—— 1994. Conodonts from the Pusku Quarry, Lower Llandovery, Estonia. *Proceedings of the Estonian Academy of Sciences Geology*, **43**, 183–191.

—— 1998. Evolution and taxonomy of the Silurian conodont *Pterospathodus*. *Palaeontology*, **41**, 1001–1050.

—— 2002. Conodonts in the Silurian of Severnaya Zemlya and Sedov archipelagos (Russia) with special reference to the genus *Ozarkodina* Branson & Mehl, 1933. *Geodiversitas*, **24**, 77–97.

—— 2007. An updated Telychian (Late Llandovery, Silurian) conodont zonation based on Baltic faunas. *Lethaia*, **40**, 45–60.

—— and ALDRIDGE, R. J. 1989. Evolution, taxonomy and relationships of the Silurian conodont *Pterospathodus*. *Palaeontology*, **32**, 893–906.

MASHKOVA, T. V. 1977. New conodonts of the *amorphognathoides* zone from the Lower Silurian of Podolia. *Paleontologicheskiy Zhournal*, **1977**, **4**, 127–131. [In Russian].

McCRACKEN, A. D. 1991a. Taxonomy and biostratigraphy of Llandovery (Silurian) conodonts in the Canadian Cordillera, northern Yukon Territory. *Geological Survey of Canada, Bulletin*, **417**, 65–95.

—— 1991b. Silurian conodont biostratigraphy of the Canadian Cordillera with a description of new Llandovery species. *Geological Survey of Canada, Bulletin*, **417**, 97–127.

—— and BARNES, C. R. 1981. Conodont biostratigraphy and paleoecology of the Ellis Bay Formation, Anticosti Island, Quebec, with special reference to the Late Ordovician – Early Silurian chronostratigraphy and the systemic boundary. *Geological Survey of Canada Bulletin*, **329**, 51–134.

——NOWLAN, G. S. and BARNES, C. R. 1980. *Gamachignathus*, a new multielement conodont genus from the latest Ordovician, Anticosti Island, Québec. *Geological Survey of Canada Paper*, **80-1C**, 103–112.

MELNIKOV, S. V. 1999. *Ordovician and Silurian conodonts from the Timan-northern Ural region.* VSEGEI, St. Petersburg, 136 pp. [In Russian].

MILLER, C. G. 1995. Ostracode and conodont distribution across the Ludlow/Přídolí boundary of Wales and the Welsh Borderland. *Palaeontology*, **38**, 341–384.

—— and ALDRIDGE, R. J. 1993. The taxonomy and apparatus structure of the Silurian distomodontid conodont *Coryssognathus* Link & Druce, 1972. *Journal of Micropalaeontology*, **12**, 241–255.

MOSTLER, H. 1967. Conodonten aus dem tieferen Silur der Kitzbühler Alpen (Tirol). *Annalen des Naturhistorischen Museums in Wien*, **71**, 295–303.

MU EN-ZHI and CHEN TING-EN 1984. New information on the Silurian strata of southern Xizang (Tibet). *Journal of Stratigraphy*, **8**, 49–55. [In Chinese].

MURPHY, M. A. and MATTI, J. C. 1983 [imprint 1982]. Lower Devonian conodonts (*hesperius* and *kindlei* Zones), central Nevada. *University of California Publications Geological Sciences*, **123**, 1–82.

——VALENZUELA-RIOS, J. I. and CARLS, P. 2004. On classification of Přídolí (Silurian)-Lochkovian (Devonian) Spathognathodontidae (conodonts). *University of California, Riverside Campus Museum Contribution*, **6**, 1–25.

NAKREM, H. A. 1986. Llandovery conodonts from the Oslo Region, Norway. *Norsk Geologisk Tidsskrift*, **66**, 121–133.

NI SHI-ZHAO 1987. Silurian conodonts. 179–183, 386–447. *In* WANG XIAO-FENG, NI SHI-ZHAO, ZENG QING-LUAN, XU GUANG-HONG, ZHOU TIAN-MEI, LI ZHI-HONG, XIANG LI-WEN and LAI CAI-GEN. *Biostratigraphy of the Yangtze Gorge area (2) Early Palaeozoic Era.* Geological Publishing House, Beijing, 641 pp. [In Chinese with English abstract].

NI YU-NAN, CHEN TING-EN, CAI CHONG-YANG, LI GUO-HUA, DUAN YAN-XUE and WANG JU-DE 1982. The Silurian rocks in Western Yunnan. *Acta Palaeontologica Sinica*, **21**, 119–132. [In Chinese with English abstract].

NICOLL, R. S. and REXROAD, C. B. 1969 [imprint 1968]. Stratigraphy and conodont paleontology of the Salamonie Dolomite and Lee Creek Member of the Brassfield Limestone (Silurian) in south-eastern Indiana and adjacent Kentucky. *Bulletin Indiana Geological Survey*, **40**, 1–73.

ORCHARD, M. J. 1980. Upper Ordovician conodonts from England and Wales. *Geologica et Palaeontologica*, **14**, 9–44.

OVER, D. J. and CHATTERTON, B. D. E. 1987. Silurian conodonts from the southern Mackenzie Mountains, Northwest Territories, Canada. *Geologica et Palaeontologica*, **21**, 1–49.

PAN JIANG 1987. Note on Silurian vertebrates of China. *Bulletin of the Chinese Academy of Geological Sciences*, **15**, 161–190. [In Chinese with English abstract].

PANDER, C. H. 1856. *Monographie der fossilen Fische des silurischen Systems der russisch-baltischen Gouvernements.* Akademie der Wissenschaften, St Petersburg, 91 pp.

POLLOCK, C. A., REXROAD, C. B. and NICOLL, R. S. 1970. Lower Silurian conodonts from northern Michigan and Ontario. *Journal of Paleontology*, **44**, 743–764.

PURNELL, M. A., DONOGHUE, P. C. J. and ALDRIDGE, R. J. 2000. Orientation and anatomical notation in conodonts. *Journal of Paleontology*, **74**, 113–122.

QIU HONG-RONG 1985. Silurian conodonts in Xizang (Tibet). *Bulletin of the Institute of Geology of the Chinese Academy of Geological Science*, **11**, 23–38. [In Chinese with English summary].

—— 1988. Early Palaeozoic conodont biostratigraphy of Xizang (Tibet). *Professional Papers on Stratigraphy and Palaeontology*, **19**, 185–202. Geological Publishing House, Beijing, [In Chinese with English abstract].

QIU JIN-YU 1990. Llandovery bioherms of Guangyuan (NW Sichuan) – Ningqiang. *Acta Palaeontologica Sinica*, **29**, 557–566. [In Chinese with English summary].

REPETSKI, J. E., PURNELL, M. A. and BARRETT, S. F. 1998. The apparatus architecture of *Phragmodus*. 91–92. *In* BAGNOLI, G. (ed.). *Seventh International Conodont Symposium held in Europe (ECOS VII), Abstracts.* Tipografia Compositori, Bologna, 129 pp.

REXROAD, C. B. 1967. Stratigraphy and conodont paleontology of the Brassfield (Silurian) in the Cincinnati Arch area. *Indiana Geological Survey, Bulletin*, **36**, 64 pp.

RHODES, F. H. T. 1953. Some British Lower Palaeozoic conodont faunas. *Philosophical Transactions of the Royal Society of London, Series B*, **237**, 261–334.

ROBISON, R. A. (ed.) 1981. *Treatise on Invertebrate Paleontology, Part W, Miscellanea, Supplement 2, Conodonta.* Geological Society of America, Boulder, Colorado and the University of Kansas Press, Lawrence, Kansas, xxviii + 202 pp.

RONG JIA-YU 1979. The *Hirnantia* fauna of China with comments on the Ordovician-Silurian boundary. *Acta Stratigraphica Sinica*, **3**, 1–29. [In Chinese].

——CHEN XU, SU YANG-ZHENG, NI YU-NAN, ZHAN REN-BIN, CHEN TING-EN, FU LI-PU, LI RONG-YU and FAN JUN-XUAN 2003. Silurian paleogeography of China. *New York State Museum Bulletin*, **493**, 243–298.

—— JOHNSON, M. E. and YANG XUE-CHANG 1984. Early Silurian (Llandovery) sea-level changes in the Upper Yangtze region of central and southwestern China. *Acta Palaeontologica Sinica*, **23**, 672–694. [In Chinese, with English summary].

SANSOM, I. J. 1996. *Pseudooneotodus*: a histological study of an Ordovician to Devonian vertebrate lineage. *Zoological Journal of the Linnean Society*, **118**, 47–57.

——ARMSTRONG, H. A. and SMITH, M. P. 1994. The apparatus architecture of *Panderodus* and its implications for coniform conodont classification. *Palaeontology*, **37**, 781–799.

SAVAGE, N. M. 1985. Silurian (Llandovery–Wenlock) conodonts from the base of the Heceta Limestone, southeastern Alaska. *Canadian Journal of Earth Sciences*, **22**, 711–727.

SCHÖNLAUB, H. P. 1971. Zur Problematik der Conodonten-Chronologie an der Wende Ordoviz/Silur mit besonderer Berücksichtigung der Verhältnisse im Llandovery. *Geologica et Palaeontologica*, **5**, 35–37.

SERPAGLI, E. 1967. I conodonti dell'Ordoviciano superiore (Ashgilliano) delle Alpi Carniche. *Bolletino della Società Paleontologica Italiana*, **6**, 30–111.

SIMPSON, A. J. 1999. Early Silurian conodonts from the Quinton Formation of the Broken River region (north-eastern Australia). *Abhandlungen der Geologischen Bundesanstalt*, **54**, 181–199.

—— and TALENT, J. A. 1995. Silurian conodonts from the headwaters of the Indi (upper Murray) and Buchan rivers, southeastern Australia, and their implications. 79–215. *In* MAWSON, R. and TALENT, J. (eds). Contributions to the First Australian Conodont Symposium (AUSCOS 1) held in Sydney Australia, 18–21 July 1995. *Courier Forschungs-Institut Senckenberg*, **182**, 573 pp.

STAUFFER, C. R. 1930. Conodonts from the Decorah Shale. *Journal of Paleontology*, **4**, 121–128.

—— 1940. Conodonts from the Devonian and associated clays of Minnesota. *Journal of Paleontology*, **14**, 417–435.

SWEET, W. C. 1979. Late Ordovician conodonts and biostratigraphy of the Western Midcontinent Province. *Brigham Young University, Geological Studies*, **26**, 45–85.

—— 1981. Macromorphology of elements and apparatuses. W5–W20. *In* ROBISON, R. A. (ed.). *Treatise on Invertebrate Paleontology, Part W, Miscellanea, Supplement 2, Conodonta*. Geological Society of America, Boulder, Colorado and the University of Kansas Press, Lawrence, Kansas, xxviii + 202 pp.

—— 1988. *The Conodonta: morphology, taxonomy, paleoecology and evolutionary history of a long-extinct animal phylum*. Clarendon Press, Oxford, x + 212 pp.

—— and SCHÖNLAUB, H. P. 1975. Conodonts of the Genus *Oulodus* Branson & Mehl, 1933. *Geologica et Palaeontologica*, **9**, 41–59.

—— THOMPSON, T. L. and SATTERFIELD, I. R. 1975. Conodont stratigraphy of the Cape Limestone (Maysvillian) of eastern Missouri. *Studies in Stratigraphy: Missouri Geological Survey, Department of Natural Resources, Report of Investigations*, **57**, 1–60.

TANG XUE-CHUN, DONG ZHI-ZHONG and QIN DE-HOU 1982. Lower Devonian of the Baoshan area, western Yunnan, and the boundary between Silurian and Devonian systems. *Journal of Stratigraphy*, **6**, 199–208. [In Chinese].

THERON, J. N. and KOVÁCS-ENDRÖDY, E. 1986. Preliminary note and description of the earliest known vascular plant, or an ancestor of vascular plants, in the flora of the Lower Silurian Cedarberg Formation, Table Mountain Group, South Africa. *South African Journal of Science*, **82**, 102–105.

TSYGANKO, V. S. and CHERMNYH, V. A. (eds) 1987. *Upper Ordovician and Lower Silurian type sections in Subpolar*

Urals. Academy of Science of SSSR, Komi Filial, Syktyvkar, 108 pp.

UYENO, T. T. and BARNES, C. R. 1981. A summary of Lower Silurian conodont biostratigraphy of the Jupiter and Chicotte formations, Anticosti Island, Québec. 173–184. *In* LESPÉRANCE, P. J. (ed.). *IUGS Subcommission on Silurian Stratigraphy (Ordovician-Silurian Boundary Working Group) Field Meeting. Anticosti-Gaspé, Québec, 1981, Volume 2, Stratigraphy and paleontology*. Département de Géologie, Université de Montréal, v + 321 pp.

—— —— 1983. Conodonts of the Jupiter and Chicotte Formations (Lower Silurian), Anticosti Island, Québec. *Geological Survey of Canada Bulletin*, **355**, 1–49.

VAN DEN BOOGAARD, M. 1990. A Ludlow conodont fauna from Irian Jaya (Indonesia). *Scripta Geologica*, **92**, 1–27.

VIIRA, V. 1982. Shallow water conodont *Ctenognathodus murchisoni* (Late Wenlock, Estonia). 63–83. *In* KALJO, D. and KLAAMANN, E. (eds). *Communities and biozones in the Baltic Silurian*. Valgus, Tallinn, 140 pp. [In Russian with English summary].

—— and ALDRIDGE, R. J. 1998. Upper Wenlock to Lower Přídolí (Silurian) conodont biostratigraphy of Saaremaa, Estonia, and a correlation with Britain. *Journal of Micropalaeontology*, **17**, 33–50.

—— —— and CURTIS, S. 2006. Conodonts of the Kivioli Member, Viivikonna Formation (Upper Ordovician) in the Kohtla section, Estonia. *Proceedings of the Estonian Academy of Sciences, Geology*, **55**, 213–240.

—— and EINASTO, R. 2003. Wenlock-Ludlow boundary beds and conodonts of Saaremaa Island, Estonia. *Proceedings of the Estonian Academy of Sciences, Geology*, **52**, 213–238.

WALLISER, O. H. 1964. Conodonten des Silurs. *Abhandlungen des Hessischen Landesamtes für Bodenforschung*, **41**, 1–106.

—— 1972. Conodont apparatuses in the Silurian. *Geologica et Palaeontologica*, **SB1**, 75–80.

—— and WANG CHENG-YUAN 1989. Upper Silurian stratigraphy and conodonts from the Qujing District, East Yunnan, China. *Courier Forschungsinstitut Senckenberg*, **110**, 111–121.

WAN ZHENG-QUAN, JIN CHUN-TAI, CHEN JI-RONG, QIAN YONG-ZHEN and YE SHAO-HUA 1991. Discovery of Late Silurian strata in the Guangyuan area of Sichuan and its significance. *Journal of Stratigraphy*, **15**, 52–55. [In Chinese].

WANG CHENG-YUAN 1980. Upper Silurian conodonts from Qujing District, Yunnan. *Acta Palaeontologica Sinica*, **19**, 369–378. [In Chinese with English abstract].

—— 1981. New observation on the age of the Yulongsi Formation of Qujing, Yunnan. *Journal of Stratigraphy*, **5**, 196. [In Chinese].

—— 1982. Upper Silurian and Lower Devonian conodonts from Lijiang, Yunnan. *Acta Palaeontologica Sinica*, **21**, 436–448. [In Chinese with English abstract].

—— 1990. Conodont biostratigraphy of China. *Courier Forschungsinstitut Senckenberg*, **118**, 591–610.

—— 1993. *Conodonts of Lower Yangtze Valley – indexes to biostratigraphy and organic metamorphic maturity*. Science Press, Beijing, 386 pp. [In Chinese with English abstract].

—— 1998. Palaeozoic conodonts from Northwest Qiangtang and Karakorum region. 343–365. *In* WEN SHI-XUAN (ed.). *Palaeontology of the Karakorum-Kunlun Mountains*. Science Press, Beijing, 365 pp. [In Chinese with English abstract].

—— 2001. Age of the Guandi Formation in Qujing District, E. Yunnan. *Journal of Stratigraphy*, **25**, 125–127. [In Chinese with English abstract].

—— and ALDRIDGE, R. J. 1996. Conodonts. 46–55. *In* CHEN XU and RONG JIA-YU (eds). *Telychian (Llandovery) of the Yangtze region and its correlation with British Isles*. Science Press, Beijing, i + 162 pp.[In Chinese with English abstract].

—— —— 1998. Comments on Silurian conodont genera proposed in Chinese literature. *Acta Micropalaeontologica Sinica*, **15**, 95–100. [In Chinese with English abstract].

—— and JEPPSSON, L. 1994. Jeppsson's ocean model and its application to Early Silurian (Llandovery) of South China Platform. *Acta Micropalaeontologica Sinica*, **11**, 71–85.

—— and WANG ZHI-HAO 1981. Conodont sequences from Cambrian to Triassic in China. *Selected papers of 12th annual meeting of the Palaeontological Society of China*, 105–115.[In Chinese with English abstract].

—— —— 1983. Review of conodont biostratigraphy in China. *Fossils and Strata*, **15**, 19–33.

—— and ZIEGLER, W. 1983. Conodonten aus Tibet. *Neues Jahrbuch fur Geologie und Paläontologie, Monatshefte*, **1983**, 69–79.

—— CHEN LI-DE, WANG YI and TANG PENG 2010. Affirmation of *Pterospathodus eopennatus* Zone (Conodonta) and the age of the Silurian Shamao Formation in Zigui, Hubei as well as the correlation of the related strata. *Acta Palaeontologica Sinica*, **49**, 10–28. [In Chinese with English abstract].

—— QU YONG-GUI, ZHANG SHU-QI, ZHENG CHUN-ZI and WANG YONG-SHENG 2004. Late Ordovician-Silurian conodonts from the Xainza (Shengzha) County, North Tibet, China. *Acta Micropalaeontologica Sinica*, **21**, 237–249. [In Chinese with English abstract].

—— WANG PING, YANG GUANG-HUA and XIE WEI 2009. Restudy on the Silurian conodont biostratigraphy of the Baizitian section in the Yanbian County, Sichuan. *Journal of Stratigraphy*, **33**, 302–317. [In Chinese with English abstract].

WANG GEN-XIAN, GENG LIANG-YU, XIAO YAO-HAI and ZUO ZHI-BI 1988. Geological age and depositional feature of the upper Xiushan Formation and Xiaoxiyu Formation of west Hunan. *Journal of Stratigraphy*, **12**, 216–225. [In Chinese].

WANG PING 2004. Conodont biostratigraphy of the Baoerhantu section in Darhan Muming'an Joint Banner, Inner Mongolia. *Acta Micropalaeontologica Sinica*, **21**, 322–331. [In Chinese with English abstract].

—— 2005. Restudy of the Palaeozoic Bateaobao section in Inner Mongolia. *Acta Micropalaeontologica Sinica*, **22**, 269–275. [In Chinese with English abstract].

WANG XIAO-FENG 1965. On the discovery of late Early and Middle Silurian graptolites from N. Kueichou and its significance.. *Acta Palaeontologica Sinica*, **13**, 118–132. [In Chinese with English abstract].

XIA FENG-SHENG 1993. Silurian conodonts from the lower part of the *amorphognathoides* Zone of the Hanjiga Mountain (north of Syram Lake), northern Xinjiang. *Acta Palaeontologica Sinica*, **32**, 197–217. [In Chinese with English abstract].

YE SHAO-HUA, JIN CHUN-TAI, HE YUAN-XIANG and WAN ZHENG-QUAN 1983. Silurian stratigraphy of the Daguan area, northeast Yunnan. *Bulletin of Chengdu Institute of Geology and Mineral Resources, Chinese Academy of Geological Sciences*, **4**, 119–140. [In Chinese].

YU HONG-JIN 1985. Conodont biostratigraphy of Middle-Upper Silurian from Xainza, northern Xizang (Tibet). *Contribution to the Geology of the Qinghai-Xizang (Tibet) Plateau*, **16**, 15–31. [In Chinese with English abstract].

ZHANG SHI-BEN and WANG CHENG-YUAN 1995. Conodont based age of the Yimugantawu Formation (Silurian). *Journal of Stratigraphy*, **19**, 133–135. [In Chinese with English abstract].

ZHANG SHUN-XIN and BARNES C. R. 2000. *Anticostiodus*, a new multielement conodont genus from the Lower Silurian, Anticosti Island, Quebec. *Journal of Paleontology*, **74**, 662–669.

—— —— 2002. A new Llandovery (Early Silurian) conodont biozonation and conodonts from the Becscie, Merrimack, and Gun River Formations, Anticosti Island, Québec. *Journal of Paleontology*, **76**, Supplement to No. 2, 46 pp.

ZHOU XI-YUN 1980. The coloration of the Silurian conodonts from Guizhou and its significance for petroleum geology. *Experimental Petroleum Geology*, **1980(3)**, 48–53. [In Chinese].

—— 1983. Colour contrast of Silurian conodonts and evaluation of hydrocarbons in Sichuan. *Natural Gas Industry*, **3**, 13–17. [In Chinese].

—— 1986. A preliminary study of the relation between the Silurian conodont distribution and the sedimentary environment of Guizhou. *Geology of Guizhou*, **3**, 423–432. [In Chinese with English abstract].

—— and YU KAI-FU 1984. Discovery of the early Silurian conodonts from Nanjiang, Chengkou, Yuechi in Sichuan. *Journal of Stratigraphy*, **8**, 67–70. [In Chinese].

—— and ZHAI ZHI-QIANG 1983. Silurian conodonts. 267–301. *In* CHENGDU INSTITUTE OF GEOLOGY AND MINERAL RESOURCES (ed.). *Paleontological Atlas of Southwest China: Volume of Microfossils*. Geological Publishing House, Beijing. 802 pp. [In Chinese].

—— QIAN YONG-ZHEN and YU HONG-JIN 1985. General description of conodont biostratigraphy of Silurian System in south-western China. *Journal of the Guizhou Institute of Technology*, **14**, 31–42. [In Chinese with English abstract].

—— ZHAI ZHI-QIANG and XIAN SI-YUAN 1981. On the Silurian conodont biostratigraphy, new genera and species in Guizhou Province. *Oil and Gas Geology*, **2**, 123–140. [In Chinese with English abstract].

ZIEGLER, A. M. and McKERROW, W. S. 1975. Silurian marine red beds. *American Journal of Science*, **275**, 31–56.

ZUO ZI-BI 1987. Discovery of Silurian conodont zoocoenosis in the northwestern Hunan and its significance in petroleum geology. *Hunan Geology*, **6**, 56–65. [In Chinese with English abstract].

APPENDIX

TABLE 1. Conodont elements recovered from the Kuanyinchiao Bed, Leijiatun Section, Shiqian County, Guizhou Province, China.

		Shiqian-2	Shiqian-1
Sample weight		2.7 kg	2.9 kg
Drepanoistodus sp.		1	
Ozarkodina aff. *hassi*	P_1	8	5
	P_2	6	5
	M	2	1
	S_0	2	1
	S_{1-2}	1	2
	S_{3-4}	2	1
Ozarkodina? sp	S_{1-2}		1
Pseudolonchodina sp.	Pb		2
	Sb?		1
	Sc		1
Walliserodus sp. A	P		5
	M		1
	Sa		4
	Sb		6
	Sc		1
Walliserodus sp.	P	1	
	Sb	1	
	Sc	1	
Wurmiella? sp.	P_1		2
	M	1	
Total conodont elements		26	39

TABLE 2. Conodont elements recovered from the Xiangshuyuan Formation, Leijiatun Section, Shiqian County, Guizhou Province, China.

		Shiqian 3	Shiqian 4	Shiqian 5	Shiqian 6
Sample weight		2.8 kg	2.3 kg	2.8 kg	2.7 kg
Coryssognathus? sp.	Sa/b		5		
	Sc			2	
	Coniform		8	7	
Decoriconus fragilis	Sb?	1			
Distomodus aff. *kentuckyensis*	Pa	1			
	Pb	1			
	M	1			
	Sb	1			
	Sc	1			
Distomodus sp.	Pb		2	1	
	Pc		3	1	
	M		3	2	
	Sa		1		
	Sc		5		
Galerodus macroexcavatus	Pa			41	
	Pc?			28	
	M			19	
	Sa			7	
	Sb			27	
	Sc			30	

TABLE 2. (*Continued*).

		Shiqian 3	Shiqian 4	Shiqian 5	Shiqian 6
Galerodus sp.	Pa		4		
	Pc?		6		
	Sb		4		
	Sc		2		
Oulodus aff. *panuarensis*	Pb	2			
	M	1			
	Sa	2			
	Sb	2			
	Sc	2			
Oulodus sp.	Pa			1	
	Pb			7	
	M			2	
	Sb			2	
	Sc			3	
Ozarkodina obesa	P_1		1	49	1
	P_2		1	10	
	M		2	6	
	S_0				
	S_{1-2}		1	5	
	S_{3-4}			7	
Ozarkodina aff. *waugoolaensis*	P_1				3
Ozarkodina sp.	M				1
	S_{3-4}				1
Panderodus panderi	pf			12	
	pt			7	
	qa			30	
	qg			29	
	qt			7	
Panderodus serratus	pf		9		
	qa		7		
	qg		13		
	qt		1		
Panderodus unicostatus	pf	8	1	150	1
	pt			69	
	qa	8		147	
	qg	3	3	230	1
	qt	1		46	
	ae			4	
Pseudobelodella spatha	S		1		
Pseudolonchodina fluegeli	Pa			12	
	Pb			4	
	M			11	
	Sa			1	
	Sb			1	
	Sc			6	
Pseudooneotodus beckmanni				2	
Rexroadus sp. A	Pa			1	
Walliserodus curvatus	P			42	1
	M	1		23	1
	Sa			29	1
	Sb	1		24	1
	Sc			17	
Walliserodus sp.	P		2		
	Sb				1
	Sc				

TABLE 2. (*Continued*).

		Shiqian 3	Shiqian 4	Shiqian 5	Shiqian 6
Wurmiella? sp.	P_1		1		
gen. et. sp. indet (oistodontan)		1			
Total conodont elements		38	86	1161	13

TABLE 3. Conodont elements recovered from the Leijiatun Formation, Leijiatun Section, Shiqian County, Guizhou Province, China.

		Shiqian 7	Shiqian 8	Shiqian 9	Shiqian 10
Sample weight		2.8 kg	2.3 kg	3.0 kg	3.0 kg
Chenodontos makros	Pa		1		
	Pb		2		
	M		4	1	
	?Sa		2		
	Sb		10	3	
	Sc		1	3	
Distomodus sp.	Pa	1			
	Pb	1			
	M	1			
Distomodus? sp.	Sb		1		
Galerodus macroexcavatus	Pa	33	21	1	
	Pc?	2	10	3	
	M	17	13	2	
	Sa	5		1	
	Sb	14	8	4	
	Sc	14	12	3	
Oulodus shiqianensis	Pb			1	
	M			1	
	Sa			1	
	Sb			2	
	Sc			2	
Oulodus sp. A	Pb	1			
	Sc	1			
Oulodus spp.	Pa	1	1		
	Pb	2			
	M		2		
	Sa		2		1
	Sb	2	3		
	Sc	3			
Ozarkodina obesa	P_1			1	
Ozarkodina parahassi	P_1			2	
Ozarkodina cf. *parainclinata*	P_1			6	
	P_2		1	8	
	M		1	8	
	S_0			3	
	S_{1-2}			4	
	S_{3-4}		1	6	
Ozarkodina pirata	P_1		21	59	
	P_2		11	12	
	M			5	
	S_0			6	
	S_{1-2}			6	
	S_{3-4}		3	12	

TABLE 3. (*Continued*).

		Shiqian 7	Shiqian 8	Shiqian 9	Shiqian 10
Ozarkodina wangzhunia	P_1	5	14		
	P_2	3	26		
	M	1	5		
	S_0		1		
	S_{1-2}	3	2		
	S_{3-4}	3	7		
Panderodus serratus	pf	24			
	pt	2			
	qa	8			
	qg	44			
	qt	4			
Panderodus unicostatus	pf		38	5	
	pt		10	3	1
	qa		21	5	
	qg		84	12	
	qt			3	
	ae			1	
Panderodus spp.	pf	2		3	
	pt	4			
	qa	6	3		
	qg	8	3		
	ae	1			
?*Pseudolonchodina expansa*	Pb	1			
Pseudolonchodina fluegeli	Pa	3			
	M	1			
	Sa	2			
	Sb	1			
	Sc	5			
Rexroadus sp. B	Pa	1			
	?Sb	1			
	?Sc	1			
Rexroadus? sp.	Pa			1	
Walliserodus curvatus	P		12		
	M		14		
	Sa		8		1
	Sb		10		
	Sc		21		
Walliserodus aff. *curvatus*	P	14		5	
	M	9		6	
	Sa	18		1	
	Sb	10		1	
	Sc	11		2	
Total conodont elements		294	409	214	3

TABLE 4. Conodont elements recovered from the Lower member, Xiushan Formation, Leijiatun Section, Shiqian County, Guizhou Province, China.

		Shiqian 14B	Shiqian 15	Shiqian 16
Sample weight		2.8 kg	1.7 kg	1.1 kg
Ctenognathodus? qiannanensis	P_1	148		
	P_2	64		
	M	50		
	S_0	25		
	S_{1-2}	43		
	S_{3-4}	101		
Distomodus cathayensis	Pa		36	
	Pb		70	
	Pc		57	
	M		78	
	Sa		23	
	Sb		95	
	Sc		118	
	Sd		26	
Distomodus sp.	Pb			2
	Sb			1
	Sc			1
Galerodus macroexcavatus	Pa			9
	Pc?			1
	M			22
	Sb			5
	Sc			9
Oulodus shiqianensis	Pa		1	
	Pb		4	
	M		2	
	Sa		9	
	Sb		1	
	Sc		14	
Oulodus tripus	Pa	9		
	Pb	17		
	M	12		
	Sa	18		
	Sb	13		
Oulodus sp. B	Sb	1		
Oulodus sp.	Pa			4
	Pb			6
	M			5
	Sa			1
	Sb			7
	Sc			9
Ozarkodina broenlundi	P_1		2	21
Ozarkodina aff. *cadiaensis*	P_1			3
Ozarkodina guizhouensis	P_1		85	
	P_2		43	
	M		37	
	S_0		6	
	S_{1-2}		9	
	S_{3-4}		38	
Ozarkodina spp.	P_2			8
	M			1
	S_0			1
	S_{1-2}			4
	S_{3-4}			6

TABLE 4. (*Continued*).

		Shiqian 14B	Shiqian 15	Shiqian 16
Panderodus unicostatus	qg		2	
Wurmiella puskuensis	P_1			15
	P_2			4
	M			2
	S_0			10
	S_{1-2}			2
	S_{3-4}			13
Total conodont elements		501	756	172

TABLE 5. Conodont elements recovered from the Upper member, Xiushan Formation, Leijiatun Section, Shiqian County, Guizhou Province, China.

		Shiqian 17	Shiqian 18	Shiqian 20
Sample weight		2.5 kg	2.0 kg	2.8 kg
Apsidognathus aulacis	Platform	33	10	6
	Ambalodontan	20	8	
	Lyriform	5	5	21
	Astrognathodontan	3	2	
	Compressed 1	8	7	
	Compressed 2	5	2	
A.ruginosus scutatus	Platform	3		21
	Lyriform			12
Apsidognathus spp.	Ambalodontan			76
	Astrognathodontan			17
	Compressed			63
Coryssognathus shaannanensis	Pa	1	2	7
	Pb	1		67
	Pc	15	8	85
	M	6	4	127
	Sa/b	9	6	35
	Sc	21	6	149
	Coniform	23	14	393
Distomodus sp.	Pb			3
	Pc	1		7
	M	5		2
	Sa			2
	Sb	1		3
	Sc			12
	Sd			1
Distomodus? sp.	Pb			1
Galerodus macroexcavatus	Pa	63	1	
	Pc?	38		
	M	167	7	
	Sa	21		
	Sb	103	4	
	Sc	92	2	
Multicostatus dazhubaensis		7	4	6
Oulodus shiqianensis	Pa		3	1
	Pb		9	5
	M		7	6
	Sa		2	3
	Sb		12	2
	Sc		15	6

TABLE 5. (*Continued*).

		Shiqian 17	Shiqian 18	Shiqian 20
Ozarkodina paraplanussima	P_1	4		
	P_2			
	M	1		
	$?S_0$	2		
	$?S_{1-2}$	1		
Ozarkodina aff. *paraplanussima*	P_1		1	
	$?S_0$		1	
	$?S_{3-4}$		1	
Ozarkodina cf. *planussima*	P_1			1
Ozarkodina waugoolaensis	P_1		1	
Ozarkodina cf. *waugoolaensis*	P_1			23
	P_2			5
	M			31
	S_{1-2}			1
	S_{3-4}			1
Panderodus panderi	pf	14	16	49
	pt	6	3	10
	qa	6	5	5
	qg	14	23	46
	qt	2	3	2
	ae	1		3
Panderodus unicostatus	pf	31	20	335
	pt	16	4	31
	qa	24	5	134
	qg	66	37	268
	qt	3	1	22
	ae			4
Panderodus spp.	pt	1		
	qg	6		1
Pseudobelodella spatha	P?	5		
	S	30	1	
Pseudolonchodina fluegeli	Pa	34	4	2
	Pb	35	6	8
	M	45	24	8
	Sa	29	3	1
	Sb	18	2	1
	Sc	72	17	4
Pseudooneotodus beckmanni		1		
Pterospathodus eopennatus	Pa	43	48	
	Pb	46	36	
	Pc	24	27	
	M	40	11	
	Sa/b	35	19	
	Sc	7	8	
	Carniodiform	18	9	
Pterospathodus sinensis	Pa			1
	Pb			10
	Pc			2
	M			6
	Sa/b			3
Rexroadus aff. *kentuckyensis*	Pa	4		
	?Pb	4		
	?M	12	1	
	?Sa	16	1	
	?Sb	21		

TABLE 5. (*Continued*).

		Shiqian 17	Shiqian 18	Shiqian 20
	?Sc	30	1	
Tuberocostadontus sp. B	?Pa		1	37
	?Pb			19
	?Pc			6
	?M	1	1	3
	?Sb-d	2	2	87
Wurmiella aff. *recava*	P_1			98
	P_2			59
	M			52
	S_0			43
	S_{1-2}			69
	S_{3-4}			76
gen. et sp. indet. B		10		
gen. et sp. indet. C				1
Total conodont elements		1431	483	2707

TABLE 6. Conodont elements recovered from the Wangjiawan Formation, Yushitan Section, Ninqiang County, Shaanxi Province, China.

		Ningqiang 7	Ningqiang 6
Sample weight		2.7 kg	2.4 kg
Coryssognathus? sp.	Pb	1	
	Pc	1	
	M	1	
	Sb	1	
	Sc	2	
	Coniform	3	
Distomodus sp.	Sc		2
Galerodus macroexcavatus	Pa	23	1
	Pc?	3	1
	M	20	1
	Sb	13	
	Sc	11	
Oulodus sp.	Pa		1
	M		1
	Sb		1
	Sc		2
Oulodus? sp.	M	2	
	Sb	1?	
Ozarkodina broenlundi	P_1		9
	$?P_2$		2
	$?S_{1-2}$		1
	$?S_{3-4}$		2
Ozarkodina aff. *obesa*	P_1	1	
Ozarkodina cf. *planussima*	P_1	1	
Ozarkodina waugoolaensis	P_1	34	
	P_2	29	
	M	12	
	S_0	3	
	S_{1-2}	10	
	S_{3-4}	21	

TABLE 6. (*Continued*).

		Ningqiang 7	Ningqiang 6
Panderodus panderi	qg		7
Panderodus unicostatus	pf	56	21
	pt	9	
	qa	64	17
	qg	53	21
	qt	4	1
Pseudolonchodina sp.	Pb	1	
	Sa	2	
Rexroadus aff. *kentuckyensis*	Pa	7	
	?Pb	12	
	?M	15	
	?Sa	3	
	?Sb	11	
	?Sc	18	
Wurmiella puskuensis	P_1	41	13
	P_2	29	7
	M	11	3
	S_0	2	4
	S_{1-2}	11	3
	S_{3-4}	23	4
Total conodont elements		565	125

TABLE 7. Conodont elements recovered from the Ningqiang Formation, Yushitan Section, Ninqiang County, Shaanxi Province, China.

		Ning. 5	Ning. 4	Ning. 8	Ning. 3	Ning. 1	Ning. 2
Sample weight		1.6 kg	2.3 kg	2.6 kg	2.5 kg	2.4 kg	2.0 kg
Apsidognathus aulacis	Platform	4	3	3	7	15	5
	Ambalodontan	2		7	10	14	
	Lyriform				3	4	
	Astrognathodontan				2	1	
	Compressed			13	5	6	
A. ruginosus scutatus	Platform		1		2		1
	Compressed		2	1	2	1	
Apsidognathus tuberculatus	Platform		2				
Apsidognathus spp.	Platform		5		1		5
	Ambalodontan		3				9
	Lyriform		2				
	Astrognathodontan		1				
	Compressed		11				1
Chenodontos makros	M				1		
Coryssognathus shaannanensis	Pa		6				
	Pb			1			
	Pc		3				3
	M		10	5			2
	Sa/b		6	5			
	Sc		7	2			
	Coniform		3	2			4
Distomodus sp.	Pb			3		1	
	Pc	1		9		1	
	M	1		2			

TABLE 7. (*Continued*).

		Ning. 5	Ning. 4	Ning. 8	Ning. 3	Ning. 1	Ning. 2
	Sa			1		1	
	Sb			3		2	1
	Sc			5	1	9	
Galerodus macroexcavatus	Pa		91	13	20	4	
	Pc?		46	4			2
	M		130	6	18	5	
	Sa		17	1			
	Sb		89	4	1	1	
	Sc		165	4	9	1	
Multicostatus dazhubaensis			11	10			
Oulodus shiqianensis	Pa		2				
	Pb		13				
	M		10				
	Sa		2				
	Sb		12				
	Sc		24				
Oulodus sp.	Sc	1					
Ozarkodina aff. *paraplanussima*	P_1						2
	S_{1-2}						1
	S_{3-4}						1
Ozarkodina cf. *planussima*	P_1				1		
	S_{1-2}				1		
	S_{3-4}				1		
Ozarkodina sp.	P_2		1				
	M		1				
	S_{3-4}		1				
Panderodus amplicostatus	pf				18		28
	pt				6		5
	qa				7		22
	qg				18		28
	qt				3		1
Panderodus panderi	pf		31	3			1
	pt		7				2
	qa		16	1			
	qg		36	6			2
	qt		2				
Panderodus unicostatus	pf		24	7		9	1
	pt		2			1	2
	qa		31	6		3	4
	qg		30	4		10	9
	qt					1	1
	ae		1				
Panderodus sp, nov. A	pf				2?		
	pt			1	3		
	qa				2?		
	qg			1	2?		
Panderodus sp.	qg		1				
Pseudobelodella spatha	P?		5				
	S		37	3		1	1
Pseudolonchodina fluegeli	Pa		17	3	3	2	3
	Pb		11	1	2		1
	M		18	2	2	2	6
	Sa		8		1	1	2
	Sb		4				2
	Sc		18	3	2	4	7

TABLE 7. (*Continued*).

		Ning. 5	Ning. 4	Ning. 8	Ning. 3	Ning. 1	Ning. 2
Pseudolonchodina sp.	Pb		1				
	Sb		2				
	Sc		3				
Pseudooneotodus beckmanni			1				
Pt. amorphognathoides subsp. indet.	Pa				3		
Pterospathodus eopennatus	Pa		24	5	9	6	
	Pb		21	10	5	12	
	Pc		8	4	6	5	
	M		11		2	5	
	Sa/b		12	2	1	4	
	Sc		14		1	1	
	Carniodiform		21	2		1	
Pterospathodus sp.	Pb						2
	M						1
	Sc						1
Tuberocostadontus sp. B	?Pa		4				
	?Pb		2				
	?M		1	1			
	?Sa						
	?Sb-d		20	2			
Tuberocostadontus sp.	?M					1	1
	?Sb-d				1		2
Walliserodus sp.	P				2		2
Wurmiella puskuensis	P_1		10	8			1
	P_2		5	6			
	M		3	4			3
	S_0		10	6			1
	S_{1-2}		9	6			
	S_{3-4}		4	9			2
Wurmiella sp.	P_1					1	
	M				4	1	
	S_0				2		
	S_{3-4}				10	4	
gen. et sp. indet. B			4				
Total conodont elements		9	1168	210	202	141	181

TABLE 8. Conodont elements recovered from the lower part of the Shenxuanyi Member, Xuanhe Section, Guangyuan County, Sichuan Province, China.

		Xuanhe 1	Xuanhe 2	Xuanhe 3	Xuanhe 4	Xuanhe 5	Xuanhe 6
Sample weight		3.1 kg	3.2 kg	2.0 kg	2.6 kg	3.1 kg	1.6 kg
Apsidognathus aulacis	Platform	23	72	6	13		
	Ambalodontan	35	26				
	Lyriform	11	10	3			
	Astrognathodontan	3	5				
	Compressed 1	12	4				
	Compressed 2	10		3			
A. ruginosus scutatus	Platform			9	7		
	Lyriform			8			
	Astrognathodontan			1			
	Compressed 1			11	2		

TABLE 8. (*Continued*).

		Xuanhe 1	Xuanhe 2	Xuanhe 3	Xuanhe 4	Xuanhe 5	Xuanhe 6
	Compressed 2			3	1		
Apsidognathus tuberculatus	Platform		6				
Apsidognathus spp.	Platform	3			23		
	Ambalodontan			16	33		
	Lyriform				9		
	Astrognathodontan			2	8		
	Compressed					1	
Coryssognathus shaannanensis	Pa			1	1		
	Pb			7	2		
	Pc	7	1	49	10	1	
	M		3	36	10		
	Sa/b	1		24	14		
	Sc	5	2	20	13		
	Coniform	27		52	16		
Distomodus? sp.	Pb	3					
	M	1					
Galerodus macroexcavatus	Pa	25				3	1
	Pc?	11				2	1
	M	34	1			3	3
	Sa	7				1	
	Sb	12	1			7	1
	Sc	24	3			4	2
Galerodus? sp. *dazhubaensis*	Pa					1	
			1	6	10	1	
Oulodus shiqianensis	Pa	22	5				
	Pb	21	2	2			
	M	24	1	1			
	Sa	8	5	8			
	Sb	29	2				
	Sc	28	16	4			
Oulodus tripus	Sa		1				
Oulodus spp.	Pa					2	
	Pb					1	
	M						1
	Sa						1
	Sc						2
Ozarkodina aff. *cadiaensis*	P_1	2					
	P_2	2					
Ozarkodina paraplanussima	P_1	25					
	P_2	6					
	M	4					
	$?S_0$	2					
	$?S_{1-2}$	8					
	$?S_{3-4}$	1?					
Ozarkodina waugoolaensis	P_1	1					
Ozarkodina cf. *waugoolaensis*	P_1		1				
	S_{3-4}		2				
Ozarkodina sp.	P_2				2		
	S_{1-2}				1		
	S_{3-4}				1		
Panderodus amplicostatus	pf	2	2	2			
	pt		2	1			
	qa		10	2			
	qg		9	4	1		

TABLE 8. (*Continued*).

		Xuanhe 1	Xuanhe 2	Xuanhe 3	Xuanhe 4	Xuanhe 5	Xuanhe 6
Panderodus panderi	pf	89	3	9	38	3	6
	pt	24	1		1	2	2
	qa	46	1	3	20	3	
	qg	99	6	11	29	5	4
	qt	34			1	5	
	ae	4					
Panderodus unicostatus	pf	314	32	13	34	27	9
	pt	28	7	12	4	5	1
	qa	167	25	10	40	10	9
	qg	226	29	18	30	22	9
	qt	53			3	4	
	ae	1			1	3	1
Panderodus sp. nov. A	pf	14	4				
	pt	12	3				
	qa	13	2				
	qg	26	8				
	qt	6					
Panderodus sp.	qt		2	4			
Pseudobelodella spatha	P?	2	1			4	
	S	10	1	1		10	
Pseudolonchodina fluegeli	Pa	53	15			1	
	Pb	31	5			1	
	M	86	48			1	
	Sa	29	4			1	1
	Sb	23	8			1	
	Sc	55	29				
Pseudolonchodina sp. nov. A	Pa				1		
	Pb				3		
	M				3		
	Sa				6		
	Sb				3		
	Sc				17		
Pseudooneotodus beckmanni		4	7		1		
Pt.amorphognathoides aff. *lennarti*	Pa						1
	Pc						1
Pterospathodus eopennatus	Pa	1	10				
	Pb		7				
	Pc		8				
	M		7				
	Sa/b		7				
	Sc		1				
	Carniodiform		9				
Pterospathodus procerus	Pa	1					
	Pb	3					
Pterospathodus sinensis	Pa				23		
	Pb				55		
	Pc				21		
	M				20		
	Sa/b				1		
	Sc				4		
Pterospathodus sinensis?	Pa	10					
Pterospathodus sp.	Pb	33					
	Pc	31				2	
	M	48				3	
	Sa	6					

TABLE 8. (*Continued*).

		Xuanhe 1	Xuanhe 2	Xuanhe 3	Xuanhe 4	Xuanhe 5	Xuanhe 6
	Sb	13				1	
	Sc	31					
	Carniodiform	61				4	
Tuberocostadontus sp. A	?Pa	2		4	3		
	?Pb	5		3	1		
	?Pc			1			
	?M			1	2		
	?Sa			1			
	?Sb-d	52		21	4		
Tuberocostadontus sp. B	?Pa		2				
	?Pb		1				
	?Sb-d		9				
Walliserodus curvatus	P		1				
Walliserodus sp.	P				2		
	Sb				1		
Wurmiella curta	P_1				17		
	P_2				16		
	M				17		
	S_0				18		
	S_{1-2}				15		
	S_{3-4}				19		
Wurmiella recava	P_1			23			
	P_2			7			
	M			9			
	S_0			2			
	S_{1-2}			8			
	S_{3-4}			20			
Wurmiella aff. *recava*	P_1	4					
	P_2	3					
	M	6					
	S_0	8					
	S_{1-2}	10					
	S_{3-4}	35					
Wurmiella sp.	P_1				6	1	
	S_{3-4}					1	
gen. et sp. indet. A			1				
Total conodont elements		2248	497	462	657	147	56

TABLE 9. Conodont elements recovered from the upper part of the Shenxuanyi Member, Xuanhe Section, Guangyuan County, Sichuan Province, China.

		Xuanhe 7	Xuanhe 9	Xuanhe 8	Xuanhe 10	Xuanhe 11
Sample weight		3.1 kg	3.2 kg	2.0 kg	2.6 kg	3.1 kg
Apsidognathus aulacis	Platform	3	5			
	Ambalodontan	1				
	Lyriform	1				
	Compressed 1	1				
Apsidognathus spp.	Platform			5		1
	Ambalodontan			2		
	Astrognathodontan			1		
Coryssognathus shaannanensis	Pb				4	15
	Pc					7

TABLE 9. (*Continued*).

		Xuanhe 7	Xuanhe 9	Xuanhe 8	Xuanhe 10	Xuanhe 11
	M					22
	Sa/b				1	11
	Sc			1	1	16
	Coniform			2	6	2
Galerodus macroexcavatus	Pa	1	4	2	36	32
	Pc?		1	2	1	1
	M		7	5	28	26
	Sa		4	1	5	1
	Sb		8	3	8	10
	Sc	1	8	3	15	14
Oulodus shiqianensis	Pa				1	2
	Pb				4	5
	M				5	2
	Sa				3	1
	Sb				4	3
	Sc				3	4
Oulodus spp.	Pb			1		
	M		2	1		
	Sa			1		
	Sb			2		
	Sc		3	3		
Ozarkodina waugoolaensis	P_1				33	10
	P_2				22	7
	M				10	3
	S_0				1	
	S_{1-2}				12	4
	S_{3-4}				9	3
Panderodus panderi	pf			5	3	
	pt			4		
	qa			4		
	qg	2		5	21	
	qt			3	1	
Panderodus unicostatus	pf	5	16	21	73	22
	pt	1	2	7	8	2
	qa	8	15	13	30	14
	qg	14	30	25	51	6
	qt	6	6	3	8	
	ae	1	2		1	
Panderodus sp. nov. A	pf		7			
	pt		4	2		
	qa		6			
	qg		10	1		
	ae		1			
Pseudobelodella spatha	P?	1	1	1		
	S	5	10	17		
Pseudolonchodina fluegeli	Pa	1	6	7		1
	Pb		2	4		
	M		2	4		1
	Sa					2
	Sb	1	1	1		
	Sc	2	8	7		2
Pterospathodus sinensis	Pa		1			
	Pb		4			
	Pc		4			
	M		4			

TABLE 9. (*Continued*).

		Xuanhe 7	Xuanhe 9	Xuanhe 8	Xuanhe 10	Xuanhe 11
	Sa/b		3			
	?Carniodiform		3			
Pterospathodus spp.	Pa			2		
	Pb	1		1		
	Pc	1				
	M			6		
	Sa/b			4		
	Carniodiform	1		2		
Tuberocostadontus sp.	?Sb-d			1		
Walliserodus sp.	P	1				
Wurmiella aff. *recava*	P_1				19	
	P_2				17	
	M				7	
	S_0				9	
	S_{1-2}				20	
	S_{3-4}				19	
Wurmiella sp.	P_1					5
	P_2					3
	M	2				
	S_0					1
	S_{1-2}	1				
	S_{3-4}	1				3
gen. et sp. indet. D				1		
Total conodont elements		63	190	186	499	264